Design Science and Innovation

The book series is intended to provide a platform for disseminating knowledge in all areas of design science and innovation, and is intended for all stakeholders in design and innovation, e.g. educators, researchers, practitioners, policy makers and students of design and innovation. With leading international experts as members of its editorial board, the series aims to disseminate knowledge that combines academic rigour and practical relevance in this area of crucial importance to the society.

Amir Shakibamanesh · Mahshid Ghorbanian

Designing Responsive City Centers

Applicable Urban Design Guidelines

Springer

Amir Shakibamanesh
Department of Urban Planning and Design
Faculty of Architecture and Urban Studies
Iran University of Art
Tehran, Iran

Mahshid Ghorbanian
Department of Landscape Architecture
School of Architecture and Environmental Design
Iran University of Science and Technology
Tehran, Iran

ISSN 2509-5986 ISSN 2509-5994 (electronic)
Design Science and Innovation
ISBN 978-981-97-8910-8 ISBN 978-981-97-8911-5 (eBook)
https://doi.org/10.1007/978-981-97-8911-5

This Springer imprint is published by the registered company Springer Nature Singapore Pte Ltd.
The registered company address is: 152 Beach Road, #21-01/04 Gateway East, Singapore 189721, Singapore

In loving memory of our fathers, Gholamreza Shakibamanesh and Ebadollah Ghorbanian, whose enduring dedication, wisdom, and unwavering love continue to inspire us every day. Their guidance and encouragement have shaped our lives and will forever light our path, even in their absence. May our humble endeavor bring joy to their souls and serve as a tribute to the profound impact they have had on us.

Preface

Throughout history, city centers have played a crucial role in the daily lives of city inhabitants, and the true character of urbanization has been realized in their context. The growth of technology and the introduction of modernism brought about a recession and altered the function of many of these centers. Thus, the subjective concept of a city's beating heart no longer exists in many surviving city centers, and their quality of life and environment decreased. New global urban development trends and related concerns and problems have a greater negative impact on metropolitan city centers than on other urban areas. Today, many cities in the world are faced with issues such as social displacements, decline in functions and activities, exclusion of economic activities, physical problems, powerlessness of critical urban spaces like city centers, over-concentration of population and activities resulting in heavy and increasing traffic, noise and air pollution, decrease in the resident population, etc. These are considered the most critical problems of city centers and have caused a decrease in the quality of urban living environments therein. However, the idea that dynamic and successful city centers could only be found in distant history is far from valid. Indeed today, a brief look at different cities reveals a plethora of quality city centers that are dynamic and attractive to the urban population.

Urban designers, planners, and city managers have for a long time focused on city centers and worked to improve the quality of life there, proposing various plans and programs to revitalize city centers in various dimensions, such as traffic, environment, tourism, economic and social factors, etc. Clearly, increasing the quality of life in city centers necessitates the development of standards and conditions that can provide inhabitants with comfort and satisfaction by addressing their material and psychological demands. As a complex and interdisciplinary concept, the design intended to improve the quality of urban life comprises both subjective and objective aspects. Therefore, topic-based or site-specific design that disregards people's beliefs, aspirations, attitudes, and dreams, as well as their unique mentality, will be ineffective in this regard.

The main focus sought in the context of city centers throughout this book is the genuine concept of "quality." The importance of this concept is to the extent that many consider it the essential duty and mission of urban design. This book extracts and infers a set of fundamental quality criteria for the urban environment regarding theories and views of various experts. It is evident that the nature of the environmental qualities of city centers, whether in the form of successful centers of the past or the form of current successful city centers, follow specific criteria. Therefore, an attempt is made to suggest requirements for realizing the obtained principles by special arrangements and urban design guidelines.

This book will result in the creation of Urban Design Guidelines by applying precise principles and well-considered arrangements to enhance the environmental quality of city centers. These urban design guidelines are given as an orderly collection of the most applicable and consistent city center design guidelines in the world, on the one hand, and the guidelines proposed by the authors of this book based on their studies and research experiences, on the other. However, these guidelines are only a small portion of a possible set of guidelines that can assist urban designers in achieving qualified responsive city centers.

The authors of this book have diligently endeavored to deliver a practical resource to the urban design community, professional leaders, and city managers, aimed at enhancing the quality of public spaces within city centers. Throughout the writing process, numerous contributors have provided invaluable insights, advice, and support, for which we are deeply grateful. We extend our heartfelt appreciation to Dr. Mostafa Behzadfar, a distinguished and insightful professor of urban design in Iran, for generously sharing his expertise and perspectives with the authors. Mr. Babak Lotfali Nejad also deserves special recognition for his artistic abilities in bringing communicative and beautiful visual expression to the all urban design guidelines described in Chap. 7 of the book, thereby facilitating the readers' comprehension of the guidelines.

We hope that this book serves as a valuable resource for urban designers, planners, and city managers, and that readers find inspiration and practical guidance within its pages to create vibrant and responsive city centers.

Tehran, Iran

Amir Shakibamanesh
Mahshid Ghorbanian

Contents

About the Authors

Dr. Amir Shakibamanesh is a distinguished Urban Designer and Associate Professor at the Iran University of Art in Tehran, Iran. With over a decade of deep expertise in the field, he has made significant strides in urban planning and design. Dr. Shakibamanesh is a prolific author, having penned more than 26 scholarly articles and twelve influential books and book chapters that explore innovative themes like smart cities, practical urban design, and the integration of virtual and augmented reality in urban settings. As the head of the Virtual Reality and Emerging Technologies Lab at the Faculty of Architecture and Urban Studies, he is at the forefront of pioneering applications that blend urban visualization, computer modeling, and emerging sensing technologies. His visionary work not only advances the discipline but also sets new benchmarks for sustainable urban development, establishing him as a pivotal figure in the future of urban design.

Dr. Mahshid Ghorbanian is an Urban Designer with more than 13 years of professional experience. She currently works as an assistant professor of urban design at the Iran University of Science and Technology, Tehran, Iran. She is the author or co-author of more than 25 scholarly articles and monographs, and author of 9 books and book chapters. Dr. Ghorbanian's research interests include urban modelling, urban spatial structure and health, morphological and visual analysis of city texture, and urban transportation in the complex city network.

List of Figures

List of Tables

Chapter 1
Urban Design Guidelines and Their Role in Enhancing Environmental Quality

1.1 Introduction

Obtaining the desired quality, which expresses itself in various functional, aesthetic, and environmental manifestations, is the primary accomplishment of an urban design project and the designer's primary objective. The pleasant feeling produced by enhancing the enumerated environmental qualities depends on a definable but wide-ranging set of criteria that can be viewed as the constituents of such qualities. An accurate and comprehensive understanding of environmental qualities and their components can assist urban design professionals in preparing guidelines to ensure their achievement. Through this crucial procedure, the urban designer will be able to convey a vast array of pleasurable and gratifying emotions through executive solutions and tangible design details. The result, which manifests the qualities and ideals of urban design in public and civic life, is referred to as urban design guidelines throughout this book.

Focusing on quality and its relationship with urban design, this chapter will discuss the views of various theorists in the field and examine the criteria for improving different qualities. As a result of this theoretical review, a comprehensive table of environmental qualities and their constituents is compiled and presented at the conclusion of the section on quality. Then, urban design guidelines and their role in urban design activities are discussed, as well as the practical considerations associated with implementing such design policies.

1.2 Quality

Improving environmental quality in urban environments is seen as the primary objective of urban design. In fact, "quality" has been discussed as the most significant accomplishment of an urban design project and its promotion as an urban designer's

A. Shakibamanesh and M. Ghorbanian, *Designing Responsive City Centers*, Design Science and Innovation, https://doi.org/10.1007/978-981-97-8911-5_1

ultimate objective. In essence, the quality of urban life is a complex and multi-dimensional concept in relationship with the conditions and status of the population in a specific geographical scale (city, region, neighborhood, etc.) that rely on qualitative and subjective factors as well as quantitative and objective ones.

Al-Qawasmi et al. (2021) also rely on the fact that there is consensus among academics that quality of life is a multifaceted construct consisting of subjective and objective dimensions that encompass a variety of facets of life experience. Urban quality of life can be described as the general well-being of people and societies living in cities and the quality of their living environment (Slavuj 2011; Al-Qawasmi 2020). The quality of urban life includes psychological dimensions such as satisfaction, happiness, and security, as well as other dimensions such as access, social opportunities, employment hopes, wealth, and leisure, are among factors that affect quality" (Seif al-Dini 2008). From this perspective, urban quality of life embodies both objective attributes (extraneous environmental and place attributes) and subjective attributes (personal insights and perception of the material and non-material conditions).

The quality of urban life is a good feeling derived from a mix of qualities related to the sense of place or identity of a city, such as legibility, collective memory, and a sense of historical belonging (Kokabi 2005). What is important here is the direct emotional connection people make to the built environment. In other words, the individuals' feelings and their internal emotional reactions to the structure of the environment and its main components are the producers of qualities.

Subjective perception of the quality of life is vital because many crucial sides of people's lives, such as the quality of the urban environment, feelings of security or social solidarity, sentimental attachment, and the quality of neighborhood relationships, are hard to measure through objective indicators (Lora et al. 2010). However, many studies have criticized the use of subjective indicators of quality of life as being unreliable and, in many cases, misleading because respondents tend to have lack of information, specific cultural manifestations of reality and biases, habitual differences, and variance in aspiration factor (Lora et al. 2010).

1.2.1 Urban Design and Its Relationship with "Quality"

A closer look at the definitions of urban design reveals that the majority of urban designers' view "improving the quality of the living environment" as the goal and pinnacle of the urban design process. Because it affects and shapes the lives of a large number of people, the quality of the urban environment and efforts to improve it are the most crucial goal of urban design. Several features, such as vitality, beauty, safety, continuity, enclosure, flexibility, character, stability, readability, quality of public territory, etc., distinguish a quality urban environment. Environmental quality refers to the desirable qualities of the environment that affect the human senses, particularly the visual, auditory, and tactile senses.

Improving environmental quality is the essential task of urban design, encompassing the design of all levels of urban projects, from designing the main structure to designing urban elements. Urban designers can use their values, concepts, tools, and methods to achieve each level's objective, thus enlivening the majority of monotonous and uninteresting urban areas.

1.2.2 Expert Opinions and Theories on Environmental Qualities

Different scholars have researched and presented the fundamental aspects and requirements for establishing a "quality physical environment" within the framework of various theories of urban design using phrases such as "responsive environments," "citizen-oriented urban planning," "good city form," etc. Diverse theories strive, despite their variations, to provide the conditions under which human needs are best addressed in public spaces, creating a pleasant and attractive environment for users of different scales (city, neighborhood, urban space, rural settlement, etc.). This chapter briefly discusses environmental quality theories and expert perspectives[1].

(A) **Kevin A. Lynch**

Lynch believes the primary purpose of urban design is to enhance the quality of the physical environment for humans, consequently enhancing their quality of life. Lynch proposes five performance factors of urban design to enhance urban quality in "Good City Form," as indicated in Table 1.1 (Lynch 1984).

Two main criteria support these five dimensions:

- Efficiency: the extent to which a place is efficient and adequate, and the costs of maintaining such values.
- Justice: balance in the distribution of environmental facilities (Lynch 1984; Patil and Patil 2016).

(B) **Francis Violich**

In Urban Reading and the Design of Small Urban Places, Violich suggests six criteria for urban design as presented in Table 1.2 (Violich 1983).

(C) **Ian Bentley**

Bentley et al. (1985) provide the qualities needed to achieve "Responsive Environments" in a book with the same title. The goal is to go through a step-by-step design approach to create a good place for people, where the artificial environment provides a democratic environment for users by enhancing the right opportunities through maximizing the range of options available to the public. Table 1.3 lists these qualities.

---

[1] Alizadeh Oskooi was the inspiration for the section on experts' and theorists' perspectives on environmental attributes (2016).

Table 1.1 Lynch's criteria for improving urban environment quality

Criteria	Definition	Subcategories	Definition
Vitality	The degree to which the form of the settlement supports the vital functions, the biological requirements and capabilities of human being, how it protects the survival of the species	Sustenance	• Access to a sufficient amount of food, fuel, water, and air, as well as waste disposal methods
		Safety	• Environmental protection from hazards, toxins, and diseases, as well as the management of such factors • Physical environment safety
		Consonance	• Environmental adaptation to the fundamental biological structure of humans
Sense	The clarity with which the city may be perceived and subjectively identified, as well as the degree to which the subjective structure of residents is consistent with societal values and concepts	Identity	The extent to which a place has an unique character that sets it apart from other places
		Structure	At the scale of a small place is the sense of how its parts fit together and in a large settlement is the sense of orientation
		Congruence	Adaptation of the structure of the environment to the activities and characteristics of the resident society
		Transparency	The degree to which one can directly witness the operations of different technical functions, activities, and social and natural processes
		Legibility	Possibility for residents to accurately communicate with one another using symbolic physical features
Fit	Compatibility between the form and capacity of spaces, pathways, and equipment, on the one hand, and the patterns and numbers of people's current and future activities, on the other	Adaptability	Compatibility between daily behavior, whether intentional or accidental, on the one hand, and the spatial context on the other
		Manipulability	A flexible space that meets basic human needs such as heat, light, dryness, accessibility, and human scale for mutual adaptation of activity and space

(continued)

Table 1.1 (continued)

Criteria	Definition	Subcategories	Definition
Access	The possibility to access to other people, activities, resources, services, information, or places	Diversity	A balanced access to other people, goods, and services
		Justice	Possibility of access to different groups in the population alongside control over the access network
		Social control	Social control over the settlement
Control	Right of access to and use of spaces and activities, and their modification, maintenance and management by users of the space	Congruence	The degree of control by actual users or inhabitants of space on it in proportion to their permanent benefits therein
		Responsibility	The degree to which people who are in charge of controlling a place have motivation, information, power, and a sense of commitment to the place and other people's needs
		Certainty	The extent to which people understand the control system, anticipate its scope, and feel secure about it

Adopted from Lynch (1984), Patil and Patil (2016)

Table 1.2 Violich's criteria for good urban design

Criteria	Definition
Readability of environment	The possibility to understand the overall system of a place
Freedom of choice	The possibility to choose various physical patterns in a place
Creating motivation through the use of heterogeneous urban forms	Availability of different contrasting physical forms in different parts of the city to increase diversity
Community livability versus private livability	A sense of friendliness and sociability in a place due to its physical form
Hearing the past sound means readability of cultural heritages;	Legibility of history and heritage of any place and search for criteria related to them
Taking into account the regional-domestic ties in the form of plans	Historical, cultural and geographical contextualism

Adopted from Violich (1983)

Table 1.3 Desirable urban design according to Bentley et al.

Criteria	Definition
Permeability	The number of alternative routes available to enter a place. There are two types of permeability (i.e., physical and visual permeability)
Variety	Variety of experiences offered by a place, which entails variety of forms, Uses, and meanings
Legibility	The quality which makes a place intelligible at two levels: physical form and activity patterns
Robustness	The possibility of a place to be used for many different purposes
Visual appropriateness	The quality related to the appearance of a place
Richness	The variety of sense-experiences which users can enjoy in a place
Personalization	Designing work and living places in a way that people can put the stamp of their own tastes and values on them

Adopted from Bentley et al. (1985)

(D) **Allan Jacobs and Donald Appleyard**

In an article titled "Toward an Urban Design Manifesto", Jacobs and Appleyard (1987) identified unfavorable settlements, extensive privatization, lack of public life, destruction of valuable places, poverty, injustice, and lack of control as contemporary design issues. Table 1.4 outlines the necessary characteristics for designing a desirable urban space (Carmona et al. 2003).

(E) **Francis Tibbalds**

In Making People Friendly Towns, Tibbalds emphasizes the significance of identifying the secret to the success and livability of some traditional cities and buildings, examining characteristics such as free pedestrian movement, place transparency, ease of access, human scale, and the integration of land uses and activities. Tibbalds argues that these characteristics should be incorporated into future constructions in order to create a "People Friendly" environment. Table 1.5 displays such qualities (Tibbalds 2012.)

(F) **Jahanshah Pakzad**

In the article "Quality Criteria for Measuring Space", Pakzad lists ten criteria for creating desirable qualities in urban spaces. Table 1.6 shows these qualities (Pakzad 2008).

(G) **DETR and CABE**

The Department of the Environment, Transport, and the Regions (DETR) and the Commission for Architecture and the Built Environment (CABE), formerly known as the Royal Fine Art Commission, have provided a more precise definition of urban design, which defines it as the art of creating a place for people. In 2000, they issued a

Table 1.4 Desirable urban design criteria suggested by Jacobs and Appleyard

Criteria	Definition
Livability	Availability of a comfortable environment free from nuisance, congestion, noise, danger, pollution, etc.
Identity and control	Creating a sense of identity in individuals and a sense of responsibility for the physical environment through facilitating engagement in urban space
Access to opportunities, imagination and joy	• Ability to comprehend a place as a cultural experience • People's access to some places to get rid of work and daily life
Authenticity and meaning	Citizens' comprehension of the meanings of a place in an engaging, inspiring, and exciting manner
Community and public life	Participation of various citizen groups in the social life of public spaces
Urban self-reliance	Self-reliance in the use of energy and other limited resources
An environment for all	Availability of urban environment for all; a place for dialogue of different values and cultures

Adopted from Jacobs and Appleyard (1987)

set of guidelines titled "By Design: Urban Design in the Planning System—Towards Better Practice", from which seven fundamental urban design quality criteria are extracted in Table 1.7.

1.2.3 Environmental Qualities and Their Constituents

Different theories and perspectives on desirable environmental features have been discussed previously. In this section, the components of environmental quality are analyzed and the recommended qualities are categorized accordingly. The quality of the urban environment is determined by three forces (components): functional quality, experiential-aesthetic quality, and environmental quality of cities.

The functional component of environment quality includes providing easy movement and access for pedestrians and riders on the one hand and entertainment, and opportunities to watch people and various ceremonies, eat, talk, read newspapers, and meet friends on the other hand, in order to ensure the vitality and richness of the city's spatial experience. Experiential-aesthetic and environmental qualities involve individuals' cognitive perceptions and environmental preferences concerning urban spaces. Finally, the "environmental component" includes topics such as micro-climate regulation of urban spaces (sunlight, airflow, shading, etc.) at

Table 1.5 Tibbalds' suggestions and criteria for people friendly environments

Criteria	Definition
Attention to public realms and their importance	A place for all Identity Contextualism Human scale
The lessons from the past	Richness Comfort Pedestrian movement Mixing uses and activities Flexibility Diversity Freedom of choice (life, work, leisure, shopping and privacy) Lasting memory Order
Mixing uses and activities	Diversity of activities Efficiency Social monitoring Social life
Human scale	Pedestrian scale (forms and activities)
Pedestrian freedom	Permeability for pedestrians Pedestrian comfort Safety Comfort Decreased pollution (visual and audio) Lack of density
Access for all	Accessibility for different groups
Clarity	Perceptibility of the environment
Lasting environments	Sustainability of places in terms of environment (historical, natural, energy, etc.)
Controlling change	Graduality of the process of urban environment change

Adopted from Tibbalds (2012)

a micro-dimensional level, while at a macro-dimensional level, it involves environmental sustainability concerns as well as the quality of balance based on the ecology of the urban environment and use of natural resources, including land, water, etc. in urban plans (Golkar 2001).

The main quality criteria of urban spaces that can be used in the development of urban design guidelines have been extracted and presented in Table 1.9 (Shakibamanesh 2008), which uses the general structure of Table 1.8 (components that make up environmental qualities) and summarizes the key parameters that make up environmental qualities taken from available literature including views of Lynch (1984), Violich (1983), Bentley et al. (1985), Jacobs and Appliard (1987), Tibbalds (2012), Pakzad (2008), and CABE & DETR (2000).

Table 1.6 Pakzad's urban design criteria

Criteria	Definition
Responsiveness to the presence of special social groups	Participation in social life for specific social groups (children, the elderly, and the disabled)
Regular and permanent use	24-h utilization of space to support and facilitate a variety of citizen activities
Responsiveness to the personality of individuals (human dignity)	Multiple dimensions of space designed to accommodate the varying demands of people, with a focus on personality and human dignity
Sense of belonging to the society and space	Creating a sense of belonging by laying the groundwork for human interaction with space and achieving harmony between the place and the behavioral patterns of users
The sense of place and nostalgia with space in individuals and groups	A unique and distinct quality for each space relative to other spaces
Involvement of different senses	Using different senses (hearing, touch, sight, and smell.) to boost the attractiveness of the urban environment
Instigation of curiosity and exploration in users	Attention to meanings conveyed by the appearance of different forms of space and its shaping elements
Efficiency of space	Ability to respond to the specific needs of people according to their cultures, use patterns, and conventions
Ease of participation	Creating the required ground for participation of people in urban space as a realm of public life
Opportunities to recall memories	Preservation of individual and collective memories in urban space changes to increase the sense of belonging to space and its impact on human beings

Adopted from Pakzad (2008)

Table 1.7 DETR and CABE's urban design criteria

Criteria	Definition
Character	A place with its own distinctive identity
Continuity and enclosure	A place where public and private spaces can be clearly distinguished
Quality of the public realm	A place with attractive and successful outdoor areas
Ease of movement	A place that is easy to arrive at and move through
Legibility	A place that has a clear image and is easily understandable
Adaptability	A place with a possibility to change easily
Diversity	A place with variety and choice

Adopted from DETR and CABE (2000)

Table 1.8 Components of urban environment quality

Functional	Quality of behavioral settings (adjustment of space, activity, and time) Quality of urban form consonance with land uses, pedestrian network, cars, etc. Quality of environmental security for activities Quality of environmental safety for activities
Experiential-aesthetic	Quality of physical-spatial environment (objective environment) including spatial structure, physical organization, mass-space, etc. Quality of perceptual-sensory environment (perceptual environment): including the quality of objective landscape, structural analysis of urban facades, etc. Quality of perceptual-subjective environment (cognitive environment): including the quality of the subjective landscape, evaluative mental landscape, vitality, etc.
Environmental	Microclimate quality of urban spaces (climatic comfort): including sunlight, shading, wind, humidity, etc. Quality of sounds, smells and aromas of the environment Quality of sustainable urban design: including efficient consumption of natural resources (energy, land, etc.), ecosystem balance and pollution reduction (water, air, and land pollutions)

Source Golkar (2001)

1.3 Urban Design Guidelines

Urban design guidelines play a crucial role in enhancing the quality of urban environments by including environmental quality variables directly into their structure and employing these elements to generate texts, visuals, and maps that assist the urban design process. They can bridge the gap between research, theory, and professional activities and are among the most effective tools accessible to urban designers for reflecting and generating quality measures and arrangements in urban spaces. Such guidelines will be able to act as general statements to define the design goals and the patterns for achieving them as well as the evidence that supports this connection between goals and patterns.

In this chapter, an attempt is made to determine the position and function of urban design guidelines in the urban design process, aiming to improve the quality of city centers and provide urban designers with a powerful tool for implementing design concepts and monitoring their implementation.

1.3.1 Urban Design Process and the Role of Design Guidelines

"Process" refers to a set of coherent and explicit acts that link the goal and the result. In other words, the activities are arranged logically and purposefully. The

Table 1.9 Environment quality criteria to be used in developing urban design guidelines

	Quality criteria	Definition
Functional	An environment for all	Accessibility of urban environment, a place to discuss different values and functional cultures
	Accessibility	Clear and easy access to other people, activities, resources, services, information, or places.
	Diversity (of activities)	Availability of various uses and activities in the space
	Flexibility	Possibility to use a place for various purposes
	Compatibility	Adaptation of the city's spatial context with uses, access network, etc.
	Community safety	Social control over the settlement by the residents or users of the spaces
	Justice	Access for different groups of the population and access network control
	Freedom of pedestrian movement	Proper infrastructure and arrangements for ease of pedestrian movement
	Gradual growth and development	Avoiding sudden changes in the urban environment
	Facilitated participation	Creating an appropriate environment for citizens in the urban space as a realm of public life
	Access to opportunities	People can break their usual frames, expand their experiences and enjoy
Experimental—Aesthetic	Social vitality (pleasantness)	A sense of a pleasant place for the participation of different groups of people in social life in the public realm
	Visual permeability	View into buildings and public spaces to increase readability and social control
	Diversity of forms and activities	Diversity of forms and activities to offer a variety of experiences
	Enclosure	Change in the sense of closure and limitation of space by modifying the height-to-width proportions of the pathways, going up and down on surfaces, covering or enclosing the pathways, etc.
	Continuity	Creating a sense of unity in space by repeating specific natural and artificial elements
	Integrity	Having a unified whole and unit

(continued)

Table 1.9 (continued)

	Quality criteria	Definition
	Comfort and convenience	Everyone should be able to feel comfortable in the city (physiologically, psychologically, etc.).
	Visual appropriateness	A quality that is related to the exterior image of a place.
	Sensory richness	Increasing the diversity of sensory experiences in a place with the possibility of using different senses
	Personalization	Designing places in a way that people are encouraged to put their own stamp on the places where they work and live.
	Legibility	Intelligibility of a place at two levels: physical form and activity patterns.
	Sense of belonging to the society and space	A sense of belonging as a result of proper grounds for human interaction with space as well as harmony between the space and the behavioral patterns of users.
	Identity (character)	A distinctive character for each space compared to other spaces.
	Opportunities to recall memories	Proper arrangements for the preservation of individual and collective memories in urban spaces.
	Contextualism	Taking into account historical and cultural aspects and using local and indigenous properties
Environmental	Geographical contextualism	Paying attention to geographical features and using the natural time and conditions of the city
	Environmental cleanliness	Preventing the pollution of natural resources and the environment
	Biological vitality	Meeting human biological needs Ensuring the survival of all organisms
	Energy efficiency	Paying attention to sustainable development in the use of resources
	Ecosystem support (ecological sustainability)	Paying attention to how to use natural resources to support the ecosystem and benefit from its natural process for blending nature in the city.
	Climate comfort	Adjusting environmental conditions such as temperature, humidity, pressure, etc. for the physiological comfort of the body

Source Shakibamanesh (2008)

concept of the urban design process or the process approach to urban design has garnered the interest of numerous urban design theorists and professionals in recent decades. The primary reasons for using the concept of process in urban design include its role in achieving better, clearer, and broader problem definitions, optimizing and considering multiple solutions, facilitating the implementation process, aiding decision-making and logical design, promoting open and explicit (non-confidential) decision-making, enhancing collective participation in design, assisting with complex decision-making, correcting decisions, evaluating and selecting alternatives (Bahreini 2020). However, the purpose of discussing the urban design process is not to address the concept of process per se but rather to examine the role of "urban design guidelines" within it. The status of design guidelines in the urban design process is shown in Fig. 1.1. As can be seen, in the third stage of the urban design process, which follows the stages of "recognizing and analyzing the environmental characteristics of the operational side of design" and "preparation of construction/development programs and plans," it is time to formulate and provide Supplemental Planning Guidance (SPG) as described in the following section.

1.3.2 The Third Stage of the Design Process: Supplementary Planning Guidance (SPG)

Local institutions can establish policies in the form of defined design concepts for specific areas and places and in connection to specific design and planning challenges by utilizing certain tools. Such instruments can be deployed by municipalities (as the most significant implementing body for urban development projects) to attain the goals stated in "Development Plan" policies.

Urban design must be considered as a policy in the Development Plan, and various approaches to achieving it must also be considered in the supplementary planning guidance (SPG), which can be particularly useful in addressing issues that (DETR and CABE 2000):

- are subject to frequent changes due to reasons beyond the control of the municipality and the city council;
- rely on recommendations and documents produced by other bodies and groups;
- are considered as advisory instruments (recommendations) rather than required ones.

Like design policies of Development Plans, SPGs are not transferrable from one place to another because they should be based on assessments of the specific conditions and grounds of each location as well as public consultation. The method of revision should be considered from the beginning of the preparation of the document. Given the need for theoretical discussion, the types of urban design policies should also be briefly mentioned before addressing SPG components.

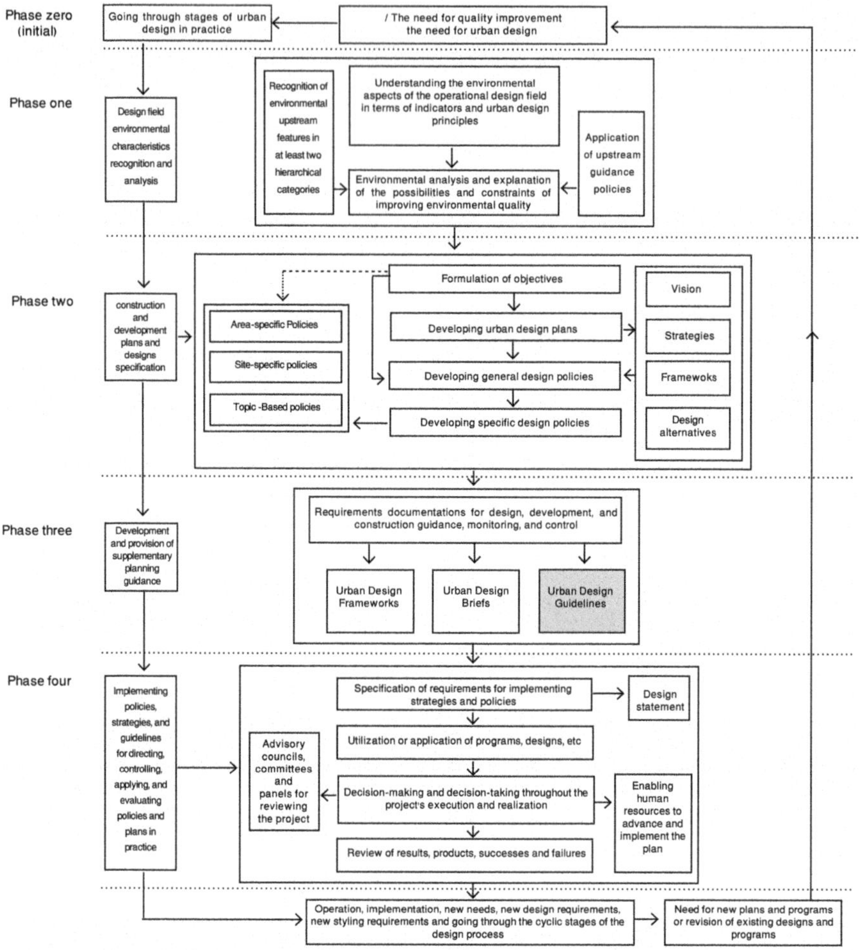

Fig. 1.1 Structure of urban design process

Supplementary Planning Guidance (SPG) is produced to provide further detail on certain policies and proposals contained within Development Plans and urban design projects. They help ensure that certain policies and proposals are better understood and applied effectively.

SPG does not have the same status or weight as adopted Development Plan policies. However, Governments typically advise that they may be taken into account as a material consideration in determining planning applications.

1.3.3 A General Classification of Urban Design Policies

Urban design policies can be classified into the following three categories based on their expected project topic, scale, and use (DETR and CABE 2000):

- area-specific;
- site-specific;
- topic-based.

(A) **Area-specific Policies**

These policies are implemented in areas that play a significant role in defining a place's distinct character and identity or in areas where change management is of particular significance. Examples of such locations include city centers, conservation areas, waterfronts, transportation corridors, regeneration zones, suburban regions, and areas having a specific function. Design policies for a particular area are developed based on evaluating its character, needs, development pressure, etc., using the information provided by design and planning professionals and local residents. If this evaluation is to be accepted as a valid basis for local design policies, it must be conducted precisely and methodically based on collaboration and consultation. Evaluation of the area's context should be analytical rather than merely descriptive.

(B) **Site-specific Policies**

Suppose the design issue is of particular importance to a particular area. In that case, the development plan should only include policies that are applicable to that area and have been adapted and produced under its conditions and characteristics.

(C) **Topic-specific Policies**

This group contains policies and methods for implementing specific arrangements and themes judged prudent by the designer. A small number of design issues may be significant enough to justify the employment of topic-specific policies and the preparation of a plan guidance instrument in the development plan.

As depicted in Fig. 1.1, supplementary planning guidance can be generated using three instruments: a) Urban Design Frameworks, b) Urban Design Briefs, and Urban Design Guidelines.

Urban design frameworks are used for the preparation of area-specific guidance documents. Design briefs are used to prepare site-specific guidance documents, and, finally, urban design guidelines are used to prepare topic-specific guidance documents.

1.3.4 Constituent Elements of Supplementary Planning Guides

A. **Urban Design Frameworks**

Urban Design Frameworks are documents that serve as a guide for places that are subject to change or that require development and modification to enhance (DETR and CABE 2000). These documents describe and illustrate how planning and design policies and principles should be implemented in an area where there is a need to control, guide, and promote change. Such areas include urban quarters, transport interchanges and corridors, regeneration areas, town centers, urban edges, housing estates, conservation areas, villages, new settlements, urban areas of special landscape value, and suburban areas identified as suitable for more intense development. The area may be likely to be developed in several phases and by several developers. It is likely to be in multiple ownerships. An urban design framework often covers an area only part of which is likely to be developed in the near future (Cabe 2003).

In reality, an Urban Design Framework is a qualitative model of development advice for changing, expanding, and forming regions. This level of "Complementary Planning Guidance" connects judgments at the strategic level with practical recommendations (Zekavat 2019.) This phase of urban design projects examines both qualitative and identity factors in the process of organization, such as spatial order, appearance, and landscape, as well as physical order, proximity, and the arrangement of both filled and empty urban spaces. It also determines the spatial arrangement of a place with respect to its surroundings. At this level, the areas, such as important roadways and urban spaces, that require a more extensive urban design investigation are outlined, as well as the scope, objectives, and function of the subsequent study. In addition, to control rules, the policy-making plan, and the design guide, the urban design framework also includes qualitative guidance and strategic development documents responsible for guiding and shaping the intervention area (Zekavat 2019).

Urban Design Framework provides a comprehensive system of design principles and policies as well as area-specific qualitative development concepts. This level of urban design research is responsible for developing a desirable model by considering all dimensions and requirements of environmental qualities, capabilities, and opportunities. An "urban development framework" often encompasses an area with limited growth potential. Using implementation strategies, such a framework goes far beyond traditional master plans. "Urban design frameworks" are known by various names, including Area Development Frameworks, Urban Design Strategies, and Planning and Urban Design Frameworks. The purpose of preparing urban design frameworks is not the setting or formulation of an inflexible structure for what should happen but to create a platform for dialogue with the builders and all stakeholders in the area (DETR and CABE 2000).

Urban design frameworks:

- explain and demonstrates the way development plan policies will be implemented in an area;
- set out a comprehensive set of design principles for the area;
- link strategy to practical suggestions;
- prepare a guidance document for development control (DETR and CABE 2000).

(b) **Urban Design Briefs**

Urban Design Briefs (UDBs) are documents that aim to provide precise information for the structural construction of urban environments on a more detailed scale than urban design studies.

UDB is a document that provides direction on developing a particular site of substantial size or sensitivity per the applicable planning and design principles. Typically, a development brief covers a property that will be substantially developed in the near future. The brief will include a suggestive, adaptable concept of future development shape (CABE 2003).

This level of urban design provides the specifics and implementation of higher-level policies for a given urban space. UDBs provide unambiguous documents for development and construction in a site and have the potential to be one of the most effective tools for elucidating the design principles as well as the implementation and viability of design policies at a site.

A development brief:

- Expresses a set of coherent development principles based on contextual analysis or planning context;
- Is prepared for a site with significant size, sensitivity and complexity;
- Minimizes ambiguities and improves the effectiveness of the planning and development process;
- Can be used to promote and enhance development (by identifying limitations, providing realistic ideas, and encouraging mixed-use and high design standards) (DETR and CABE 2000).

(C) **Urban Design Guidelines**

Urban design guideline (UDG) is a general term for SPG documents that guide developers and their designers (and other agencies) during the planning and design of the development. It can be created by municipal authorities, landowners, developers, partnerships, and corporate and community organizations (who should all be included in the process) or by a combination of these forces. The guidance should be used as a basis for discussions and negotiations with developers, and for decisions on planning applications (CABE 2003).

UDG connects the implementation of rules and regulations to general planning policies. The primary goal of design standards is to communicate the designer's preferred quality for a location. UDG is a bridge between academic research and professional practice. Alexander (1977) in Pattern Language propose the most generalized version of UDG. These guidelines are similar to general statements in that they outline the objectives, the required design patterns for accomplishing them, and the evidence demonstrating the relationship between those objectives and patterns. In coordinated urban design, the desired guidelines are those that aid in the realization of the design's intended purpose and concept. Therefore, the emphasis is on project-specific principles or design guidelines for completing the components of a coordinated urban design (Lang 2006).

Over the ages, the nature of design rules has altered little. In Italy, for instance, the history of façade design guidelines that define the nature of a new building's openings dates back to at least the thirteenth century. What has changed and will continue to change in the future are the perceptions of the mechanisms used to attain design objectives and the types of standards utilized to ensure these mechanisms are used in the design (Lang 2006).

The most effective rules are those that are backed by all authorities, including council officials, city administrators, and members of the planning committee, and are easily understood by local builders, architects, and other users. Design guidelines can inspire a design that is both imaginative and contextually appropriate. They are able to promote norms for a certain type of development whose concerns have been identified and respond to queries typically posed by urban construction permit applicants. In addition, design guidelines enable the municipality to guide development concerning particular design challenges and types of development, thereby clarifying design policies in the development plan (DETR and CABE 2000).

The efficiency and effectiveness of an urban design guideline or other Supplementary Planning Guidance (design briefs or urban design frameworks) depend on:

- The level of commitment of departments and areas related to the document;
- The level of support by council members, city officials, and city managers for urban design guidelines;
- Efficiency and effectiveness of public participation in the preparation of such guides; and
- The logical process of structuring urban design guidelines and clarity in their statement and illustration.

1.3.5 Types of Urban Design Guidelines

Specifications for open spaces, such as streets and squares, as well as the structures that enclose them, may be included in these guidelines. Patterns that a structure, complex, or a part of a building must adhere to are known as prescriptive guidelines (for example, all buildings must have red brick strips at 5-meter intervals). Functional requirements specify how one or more structures must operate. (For instance, no shade should be employed over open areas between 11 a.m. and 2 p.m. between November and February, when the sun's angle is lower than usual.) The first scenario is preferred by investors since the requisite shapes are stated without ambiguity. This volume focuses on the third group, which consists of recommended guidelines. While prescriptive and functional guidelines are legally enforceable if codified and recognized by the constitution, recommended guidelines are not (Lang 2006).

1.3.6 Contents of Urban Design Guidelines

Although there is no single global standard structure and format for setting up design guidelines, the most efficient and common ones have a similar overall structure. A well-prepared urban design guideline includes the following (DETR and CABE 2000, American Planning Association 2006):

(a) the purpose of the guideline
(b) information on how to use the guideline and who is responsible for preparing it;
(c) consultation on the guideline and municipality's response;
(d) current status (consultant's draft for example) and final status (Supplementary Urban Planning Guide, for example)
(e) Description of upper-level policies and the guide's relationship to Development Plan policies, other national guidance documents, other design guidelines, development briefs, and special relevant programs, such as those pertaining to city center administration, security, and preservation;
(f) A description of the context assessments on which the guideline is based and a summary of what such assessments have shown;
(g) An account of current design policies in the "Development Plan" tailored to specific topics or specific local contexts. In fact, like all legal provisions, support for the design guidelines must be within the framework of policies adopted by the legislature. Therefore, there should be a reference to the relevant legal policies in each section of the design guidelines. Furthermore, footnotes or endnotes can show the link between a particular design guideline and a policy.
(h) Principles: Design principles in a project must be stated as the main topics of design guideline groups. In order to satisfy the principle or principles proposed at the beginning of one or a set of urban guidelines, intents are specified. In fact, intents show that the design guidelines have been developed in line with the principle or principles and have been compiled and presented for a specific "purpose";
(i) Titles: The categories in question should be established with titles for various subjects of interest from the design guidelines.
(j) Using illustrations, photos, and plans with captions. Though the written language in design guidelines, particularly regarding legal aspects, is needed, design guidelines usually rely on photos, sketches, and diagrams to convey concepts and information. While text alone is appropriate for explaining design concepts and ideas, these design guides should be meaningful and understandable to planners, designers, and the general public. The format used in preparing

formal and legal reports on design guidelines is similar to the vision section in planning reports. If photos are used in the design guide, an attempt should be made to use, as much as possible, the buildings (or parts thereof) or elements and urban spaces from within the target community that are desirable and ideal to be imitated. In the absence of successful and good examples in the target community, the examples in the surrounding communities can be used (which, depending on the subject of the design project, can be a city, section, corridor, neighborhood, etc.). However, alternative samples provided for imitation should be available in areas with similar climatic, planting, topographic, and cultural conditions.

(k) Classification and observance of hierarchy in the compilation of design guidelines. More detailed concepts and clauses related to the main urban design guidelines can be presented in separate sections or attached to each design guide (or in both ways). This leads to further confirmation of the main urban design guidelines and shows the rational arguments behind a decision. In addition, since the guides are likely to be reviewed and revised, this organization is a reminder of the main goals and objectives of the guides. It should be noted that design guidelines are not, of course, able to cover all the topics and issues in the field of urban design, but they should nevertheless include the essential topics and titles. This procedure causes the most effective design guidelines to be considered, critiqued, evaluated, and revised in a more focused and straightforward review of design guidelines and helps avoid addressing minor and trivial issues. In addition, it informs the development committee of the issues requiring a higher degree of attention. The arrangement of critical topics in the structure of design guidelines makes the final design more visible and focuses and aligns the power of reviewers, designers, and decision-makers. Finally, this action can reduce the number of changes and successive modifications of design guidelines by specifying the most important measure to be taken in the implementation of design arrangements.

(l) The Glossary. The use of glossaries in the preparation of reports on design guidelines can aid in reducing the number of debates and contradictory perceptions resulting from varying interpretations of synonymous terminology. Some phrases, such as "compatibility," are employed in the preparation of guidelines with a broad range of definitions. Such terms should not be used directly in urban design guidelines, but they may be appropriate for inclusion in the policies or principles of a design guideline report. If "compatibility" is intended as a goal, design guidelines should attempt to specify strategies to accomplish it. In addition, the texts and writings of design guidelines should avoid ambiguous and overstated language.

(m) Sources and references for further information

Taking into account the different sections mentioned above alongside the subject of the design guide, Table 1.9 can be used to determine the principles of achievable quality. "Objectives" or "intentions" establish the relationship between the principles and guidelines of urban design and determine the purposes of setting those guidelines.

In fact, intentions ensure that the guidelines have been set in line with the principles and have been developed and presented for a specific "intention." In the seventh chapter of the present book, several examples of arranging urban design guidelines aimed at improving the quality of city centers have been compiled and presented.

1.3.7 *Major Themes for Categorizing Urban Design Guidelines*

Design guidelines address a wide range of topics pertaining to development control and design, including the following which are deemed to be the most crucial.

(A) **Architectural Design Guidelines for Buildings**

This set of design guidelines addresses topics including building facades, signs, building annexes, building appearance (details and materials), scale (massing), the unique character of buildings, design integrity, articulation of buildings, the relationship between buildings and sidewalks, empty walls, setbacks, new developments, building entrances, awnings, and guardrails, etc. For example, building character and legibility (ease of perception) are the most important objectives discussed under design guidelines for group facades.

(B) **Landscape Design Guidelines**

Pathways and spaces design guidelines address themes such as landscaping, pavements, vegetation and trees, furniture and amenities, lighting, public arts, public spaces, and the like. Promotion of character, public arena quality, ease of movement, and legibility (ease of perception) are even more important in this area. Landscape design guidelines also cover natural conservation and environmental quality improvement topics.

(C) **Neighborhood, District and City Design Guidelines**

Some design guidelines relate to areas such as neighborhoods and all or part of cities and towns. In fact, they form the macroscale of urban design guidelines and can play a valuable role in guiding development, promoting mixed-use, motivating revitalization, contributing to the city's prosperity, and acting as a vision document.

References

Al-Qawasmi J, Saeed M, Asfour OS, Aldosary AS (2021) Assessing urban quality of life: developing the criteria for Saudi cities. Front Built Environ 7:682391

Al-Qawasmi J (2020) Measuring quality of life in urban areas: toward an integrated approach. Int J Environ Sci Nat Res 25(2):67–74. https://doi.org/10.19080/IJESNR.2020.25.556158

Alexander C (1977) A pattern language: towns, buildings, construction. Oxford University Press

Alizadeh Oskoui M (2016) Citizen-oriented neighborhood design: A case study of Tabriz Street neighborhood. Master's thesis. Iran University of Science and Technology, Tehran, Iran
American Planning Association (2006) Planning and urban design standards. Wiley
Bahreini SH (2020) Urban design process, 11th edn. Tehran University Press, Tehran, Iran
Bentley I, McGlynn S, Smith G, Alcock A (1985) Responsive environments. Routledge
Carmona M, Oc T, Tiesdell S, Heath T (2003) Public places—urban spaces. Architectural Press
Commission for Architecture and the Built Environment (CABE) (2003) The councilor's guide to urban design. Retrieved from https://www.designcouncil.org.uk/our-work/skills-learning/resources/councillors-guide-urban-design/#:~:text=The%20Councillor's%20guide%20to%20urban%20design%20provides%20the%20essential%20knowledge,to%20have%20been%20involved%20in.
Golkar K (2001) Exploration in the definition of urban design, 1st edn. Center for Studies and Research of Urban Planning and Architecture of Iran
Jacobs A, Appleyard D (1987) Toward an urban design manifesto. J Am Plan Assoc 53(1):112–120
Kokabi A (2005) Planning the quality of life in urban centers, definitions and indicators. Master's thesis. Tarbiat Modares University
Lang J (2006) Urban design: A typology of procedures and products. Routledge
Lora E, Powell A, van Praag B, Sanguinetti P (2010) Preface. Lora E, Powell A, van Praag BMS, Sanguinetti P (eds). In: The quality of life in Latin American cities: Markets and perception The Inter-American Development Bank, Washington, pp xix–xxiv
Lynch K (1984) Good city form. MIT Press
Pakzad J (2008) Quality criteria for measuring space. Abadi J 39
Patil R, Patil V (2016) Theory of good city form by Kevin Lynch review. Int J Innov Res Sci Eng Technol 5(1)
Saif al-Dini F (2008) Fundamentals of urban planning. Aizh Publication
Shakibamanesh A (2008) Promoting city centers quality by using special urban design guidelines. Master's thesis. Iran University of Science and Technology, Tehran, Iran
Slavuj L (2011) Urban quality of life—A case study: the city of Rijeka. Croatian Geographical Bulletin 73(1):99–110
Tibbalds F (2012) Making people-friendly towns: improving the public environment in towns and cities. Taylor & Francis
The Department of the Environment, Transport, and the Regions (DETR) and the Commission for Architecture and the Built Environment (CABE) (2000) By design: urban design in the planning system: towards better practice. Thomas Telford
Violich F (1983) Urban reading and the design of small urban places: The village of Sutivan. Town Planning Rev 54(1):41–62
Zekavat K (2019) The position and content of urban design documents in the spatial planning system (an introduction to urban design methods and techniques). Shahid Beheshti University Press, Tehran

Chapter 2
An Introduction to the Concept of City Centers

2.1 Introduction

In this chapter, definitions and views on city centers are presented to acquaint the readers with various opinions about the city center concept. The qualities and criteria that have been present in the majority of the thriving city centers around the world are then categorized and compiled logically for use in current designs based on a thorough and meticulous review of past city centers. Such qualities and characteristics include accessibility, diversity of uses, distinctive identity of the centers (as manifested in their symbolic elements), physical compactness of the centers and their human spatial enclosure, variety of building types, and availability of open public spaces among such city centers. This chapter concludes with an overview of city centers' spatial system and spatial hierarchy.

After reading the discussions on the said topics, the reader is expected to be able to provide a clear definition of an ideal city center based on criteria derived from successful examples of city centers over time and to create a useful theoretical framework for the recreation and creation of flourishing vibrant city centers putting the suggested factors in the context of current design needs and requirements.

2.2 Definitions and Concepts of City Centers

The city center is a part of the urban territory that is typically located close to the geometric center of the city and is regarded as the most vital area by its residents. It is where the majority of important events, such as shopping, social interactions, festivals, and celebrations, take place. In almost every community, the city center is bustling with activity and vitality (Ford 2003).

Saidi Rizvani (2001), citing Frederic Gibberd, defines a city center as "a combination of administrative, business, artistic, and cultural centers for the entire city

A. Shakibamanesh and M. Ghorbanian, *Designing Responsive City Centers*,
Design Science and Innovation, https://doi.org/10.1007/978-981-97-8911-5_2

that are the manifestation of all people's social presence and interactions." Therefore, the most significant manifestations of the city should be located in the city center, offering the highest level of urbanity, architectural values, and environmental characteristics (Saidi Rizvani 2001).

The city center is an area that is almost centrally located among other districts. The city center is sometimes called the heart of the city. This city region has a high population density and can independently accommodate several administrative districts. The city center is regarded as the city's economic hub. The city center should be functionally designed for all social classes. Socially, the public space of the city center should have night uses. The city center should contain artistically significant landmarks expressing the identity of the city and its inhabitants. Additionally, climatic considerations such as covered sidewalks, ventilation, heating, and cooling must be considered. Finally, the city center should be close to the city's geographic center (Golany 1976).

In the introduction of the book entitled "Creating a Vibrant City Center", Paumier (2004) begins the discussion of city centers as follows:

> After decades of chaos and confusion in cities, the changes and transformations in the first half of the twentieth century, and the neglect and destruction of city centers, it is time again for these centers to be restored to able to establish their high identity and prestige in terms of culture, business, and most importantly, as the most important places for the emergence of collective interactions. City centers should be able to reflect the spirit and life of the city as in the past, and their diversity, color, activity, and liveliness should turn cities into places for search, discovery, entertainment, and reception, as well as for seeing and being seen, interacting, learning, and enjoying. City centers must be able to act as a safe haven and home to millions of people and a place to attract large numbers of city visitors, creating a special place to improve the social and cultural conditions and fertilize the macro-regional and micro-local economy.

According to Project for Public Spaces (PPS), "important city centers make cities exuberant and full of life." Therefore, a city can be considered healthy when they have lively city centers full of collective activities. "The process of revitalizing and strengthening city centers also accelerates the process of development of the entire city. These centers should connect people to institutions, organizations, and public spaces." (PPS, 2008). The LMN Architects Company points out that a city center is a public space that has a special personality and identity that makes it a unique destination for people and can reflect the civic character of the whole city (LMN 1998).

In fact, a city center is where people meet each other for various reasons and to have face-to-face contact. People may go to the city center for various functions such as cultural and commercial/business activities or psychosocial needs. The word city center per se has a geometric connotation in the sense that it may be located in the geometric center of the city. However, this is not necessarily the case for all city centers. Saeidnia (2004) states that the first nucleus of most cities and the city's main market gradually turn into the city center, where commercial and social activities are concentrated, and transportation lines lead from all parts of the city (Saeidnia 2004). Historic city centers of Iran, Turkey, and a few other countries in Middle

Fig. 2.1 Old Iranian bazaars, as the backbone of city centers. *Source* Public domain image, available at Wikipedia

East typically defined by bazaar which constitute the main backbone of urban core (Fig. 2.1).

Table 2.1. Presents the main terms and concepts related to the city center that can be effective in clarifying the similarities and differences of the expressions used in relation to city centers.

Given the preceding definitions and the expansive semantic and conceptual scope of the city center, it is unavoidable to focus on the practical components of the notion. Accordingly, this volume uses the following definition of the city center as the basis of study and design.

> The city center can be described as the focal and central area of the city – but not necessarily in a geometric sense – that is the context for the formation and accumulation of the most significant urban activities and the collective memory of the people. It is the manifestation of the social presence of all the city's inhabitants and the site of the majority of the city's administrative and economic interactions. In addition, the center of each city shapes its identity and character and is regarded by its citizens as the most important part of the city.

On the basis of this definition and prior research and experiences, the following are the most significant characteristics and features of the city center. A city's center:

- Is the most important or at least one of the most important urban elements in the development process of the city.
- Provides proper access from different places in the city.
- Is the main arena for social interactions and expression of collective memories of the people, as well as the symbol of social and cultural identity of the city.
- Is attractive for all groups of people.
- Is the main commercial and administrative center of the city.
- Serves as the hub of numerous activities and functions.
- Manifests the outstanding quality of an urban space as a symbol of the whole city.

Table 2.1 The main terms and concepts related to the city center

Term	Definition	Global experiments
Civic center	The word's origin can be traced to the Hellenistic concept of an acropolis and the Roman idea of a forum. The part of the city where municipal buildings and sometimes recreational and cultural centers are concentrated. The civic center is usually adjacent to a commercial center and is created and used to show the nature of a city for public activities (Kokabi 2005). Indeed, a civic center is a prominent land area within a community constructed to be its focal point or center. It usually contains one or more dominant public buildings (Wikipedia 2011). These city centers often accommodate nearby parks and plazas, central offices, a city hall, courtrooms, post offices, municipal service offices, etc.	 Brent Civic Center, London, United Kingdom; *Source* Mattbuck4950 (2013)
Central Business District (CBD)	CBD refers to the district where business and commercial activities of the city prevail. That is, commercial and related uses are of the highest density in the CBD. Moreover, building density, commercial land value, and density of pedestrian and car access are the highest in the CBD. In American cities, the CBDs were the locus of concentration of service and business activities and were often devoid of residential uses. The development in such cities was associated with the growth of the third sector of the economy (services) (Farid 2010)	 Chicago Central Business District, USA; *Source* Ken Lund (2013)

(continued)

Table 2.1 (continued)

Term	Definition	Global experiments
Community centers	Community centers fall between local shopping centers and regional centers in terms of size and variety of activities. They are about the size of a small-town center and generally provide monthly goods and services to residents. The term is also used for local congregations and refers to several buildings designated for public use, including a library, a meeting room, shops, and schools in a small town. The community center does not serve the whole city but only meets essential needs (Kokabi 2005)	 Jefferson Park Community Center, Seattle, Washington *Source* Seattle Parks and Recreation (2014)
Town center	Cowan defines town centers as 'city, town and district centers, irrespective of size, which provide a broad range of facilities and services, and which function as a focus for both the community and public transport, excluding retail parks, neighborhood centers and small parades of shops of purely local significance (Cowan 2005)	 Dijon Town Center, France *Authors' Personal Archive, 2022*
Downtown	This term was first used to refer to the uptown and downtown areas in Manhattan in the United States and was later extended to the business district of an American city. Today, it is spreading to the urban planning literature of other countries across the world (Cowan 2005). As a new understanding of their purposes has emerged over the past fifty years, the central business districts of major US cities have changed in content. Housing in these places has grown in importance and has significantly aided in shifting functions. (Birch 2009)	 Downtown Toronto, Canada *Source* Spragg & Bernard (2019)

(continued)

Table 2.1 (continued)

Term	Definition	Global experiments
City center	The city center is located at the center of some larger districts and is sometimes referred to as the heart of the city. This area of the city has a high population density and can accommodate several independent administrative districts. The city center is considered the economic center of the city. In fact, the first nucleus of most cities and the city's main market gradually turns into the city center, where commercial and social activities are concentrated and to which major transportation lines from all sides of the city end (Shakibamanesh 2008)	 Naghsh-e Jahan Square, Isfahan Historic city center, Iran; *Authors' Personal Archive, 2017*
Shopping centers	Typically, shopping centers are made up of a cluster of stores that are built next to one another. Trajan's Forum/(Market), which was constructed in the second century AD, according to Hall (1998), created a sizable shopping complex with an indoor hall housing roughly 150 businesses. However, Lake Forrest, Illinois, saw the construction of the first planned shopping complex in 1916. (Cowan 2005). Typically, shopping center buildings were one-story, low-cost structures that served only as places to buy and sell goods (Francaviglia 1996)	 Trajan's Market, Rome, Italy *Source* Wikimedia Commons (2013)

(continued)

Table 2.1 (continued)

Term	Definition	Global experiments
Mall	In the process of renovation of old city centers after World War II, and the design of new cities and their suburbs, planned commercial centers were formed, leading to large commercial-service centers for pedestrians. The first examples of this type of shopping center in the United States were built during the 1950s and 1960s. These malls could be indoor or outdoor and are often in the form of an island of shops, service agencies, and commercial offices surrounded by a sea of parking lots (Brambilla 1977)	 Shopping Malls in America *Source* Wikipedia (2008)
City core	The city core of is smaller than the city center and includes the most valuable land, the most intensive building development, and the greatest number of pedestrians and vehicles (Kokabi 2005). Typically, the city's core holds the origins of the city. Consequently, they frequently encompass the city's historic districts	 Bologna Historic City Core, Italy *Authors' Personal Archive, 2019*

- Organizes the surrounding public and collective spaces and places.

As can be seen, the significance of the city center resides in its attraction, which depends primarily on its accessibility and multi-functionality. Given the concentration and aggregation of the city's important human, social, economic, historical, cultural, and physical activity in a small area, the city center could be considered its beating heart.

2.3 Common Characteristics of City Centers in the Past

Between the eighteenth century and the middle of the twentieth century, the city center was the heart of a region's social and economic life, where people congregated to trade the commodities, they had made, meet each other, and exchange information and opinions on a variety of issues. In reality, the city center was a civic and cultural center and served as the symbol of the urban society's identity (Paumier 2004).

Even while social and economic pressures have altered the form and physical function of the city center, the same inherent characteristics of historic cities continue to be important and crucial to the development of these centers. The latest wave of urban reconstruction has consisted of efforts to recreate an environment reflective of normal urban life. Looking at the evolution of city centers, this section tries to apply previous experiences of designing and creating city centers to today's conditions.

Historically, unique characteristics introduced city centers as places for people to congregate, conduct business, shop, and reside. These features include ease of access, diversity of uses, open public spaces, formation of identity through symbolic elements, physical compactness and concentration in city centers, enclosure with human scale, and diversity and uniqueness of buildings alongside their integration in the city center.

2.3.1 *Ease of Access*

In historical cities, the city center was regarded as the gravitational core of the transportation network. As long as the city center had good access to land (roads), water, and railroads, they had a significant competitive advantage for production and commercial activities, generating a type of commercial concentration that led to substantial investments in a thriving economy. When walking and horseback riding were the primary forms of mobility, the majority of city center activities had to be located within a short distance of one another (Fig. 2.2). As homes and workplaces were required to be in close proximity, historic city centers were major residential and commercial areas, and their neighborhood units were home to people from many walks of life (Paumier 2004).

Fig. 2.2 In some city centers, you can still see tourists moving around and exploring the city in horse-drawn carriages. Italy's Florence. Authors' Personal Archive, 2019

2.3.2 Diversity in Uses

In order to be responsive to varied functions, city centers have developed a wide variety of mixed uses. The desire for proximity (easiness of access) also resulted in the dense presence of people and activities that were highly diverse and intertwined.

The ground levels of streets were the greatest locations for commercial uses. While the upper floors were frequently used for residential and other non-commercial uses, the sidewalk stores served a significant and highly visible business function, such as buying and selling. Other significant applications were city halls, civic buildings, courtrooms, post offices, and libraries. Also, there was a significant amount of office space in city centers. In city centers, other business activities such as retail stores, restaurants, hotels, banks, services, and entertainment venues such as theaters, night-clubs, and bars were frequently found (Ford 2003). There were often places for people to live in city centers. Some buildings for sale had apartments on the upper floors for residence. Hotels usually accommodated people for long periods; hence they were sometimes the permanent address of residents. Residential homes were also often built in or near the city center (Southworth 2005).

Due to the diversity of the mix of uses, the absence or inefficiency of a single component did not negatively impact the city center's economy as a whole. During the Industrial Revolution, many cities were unduly reliant on manufacturing, which limited the range of commercial uses and led to the decline of many city centers (Paumier 2004).

The combination of mixed-use buildings and dense pedestrian activity in historic city cores produced a vibrant social atmosphere. Mixed uses ensured opportunities for encounters and contacts without prior coordination and facilitated gatherings and social activities, transforming the central area into a significant hub for community interactions. Consequently, the entire city's social and economic vibrancy expanded due to the enhanced information and opinion interchange (Fig. 2.3).

Fig. 2.3 Mixed-use in vibrant urban cores from the past to present; city center area near the bazaar in Tabriz, Iran. Authors' Personal Archive, 2016

Fig. 2.4 U.S. Post Office, Custom House and Court House, Cleveland, Ohio, 1933. *Source* Library of Congress (1910)

2.3.3 Inclusion of Open Public Spaces

A few cities allocated a block to creating a plaza or an outdoor market to form a small open space in the real core of the city center. Occasionally, alterations to the street network might generate unplanned open spaces. The scarcity of open spaces boosted their value and significance as a counterpoint to the city's architecture (Paumier 2004). However, there was a public space in almost every city center. These spaces often included green spaces, Court House Square, and other types of public spaces. Eventually, religious, commercial, and government uses surrounded these squares or green spaces. These urban cores in North America resembled medieval villages rather than formally planned cities (Francaviglia 1996; Bell 2005) (Fig. 2.4).

2.3.4 Formation of Identity via Symbolic Components

Certain physical aspects and components in city centers have taken on symbolic significance. These elements include buildings with aboriginal architecture, such as clock towers and valuable landscaping. Although wise and appropriate use of such symbolic components can improve the physical quality of the urban environment, in some situations, urban designers and architects attempt to include them with more frequency and significance than their historical relevance.

Fig. 2.5 The possibility of using physical elements representing a special identity and character in the design of city centers; Tabriz, Iran. Authors' Personal Archive, 2016

In this regard, some public and private developers have gone so far as to include artificial historical components in the design of city centers to give a sense of historical heritage to a place that is in fact, new (Jacobs 2005). Today, the extreme use and mere imitation of identity-related and historical elements in city centers has been widely criticized, what Jacobs interprets as "cultural brainwashing." In this regard, as demonstrated by scholarly and professional texts in the field of urban design, designers and experts should strive to protect and preserve works of physical value in city centers and use indigenous and identity-forming patterns in the form of coordinated designs in accordance with the temporal and local requirements of the project, while avoiding blind imitations (Fig. 2.5).

2.3.5 Density and Physical Compactness of City Centers

The high demand for various uses and functions in the city center increased the value of land in these cores. As a result, multistory structures were constructed so landowners could generate greater returns from real estate investments. Typically, structures covered all or nearly all of the surface area of particular parcels of land, leaving just a small area for service areas, skylights, etc.

The combination of factors, including the pedestrian-oriented nature of the centers, the need for easy access to its various parts, and the rising value of land and buildings in the city center, naturally compacted and centralized the physical structure of these centers. Later, this congestion disappeared due to the advent of cars, the sprawling development of suburbs, and the rise of manufacturing technologies, giving way to large new commercial centers prioritizing car traffic.

2.3.6 Enclosure and Human Scale

As mentioned, the density of development in traditional city centers was high due to increased land value. Maximized site coverage was concomitant with construction near the streets, which gave an intense sense of spatial enclosure.

Although the city centers were densely developed, construction conditions and regulations limited the buildings' height, creating a human scale. Coordination of the height of buildings and massing styles strengthened the human scale of the passages and resulted in architectural harmony and visual coherence in the center (Paumier 2004).

These buildings were constructed using natural materials such as wood, brick, and stone. Typically, sidewalk views of buildings were about 6 to 9 meters wide. Each building featured at least one, and frequently two or more, entrances. These elements played a significant effect in enhancing the human scale of the city center.

2.3.7 Inclusion of Diversity and Unique Character in Building Types Without Losing Integration

Over time, city center buildings have been developed individually based on economic needs. Older cities have usually experienced several generations of buildings. Various buildings have been demolished and replaced by newer ones. In some cases, where a larger structure than the previous one was required, the new building was a combination of several pieces that previously hosted smaller buildings, leading to a wide variety of buildings in terms of size, shape, and age in city centers (Langdon 1994).

Moreover, many buildings in city centers were in distinct parts and had different owners (Francaviglia 1996), which contributed to the diversity of different opinions and tastes.

Since the street was the main pathway to access the market, the competition for the street edge was much greater than in other parts. The depth of building blocks was usually more remarkable than their width, which created a pattern of narrow views across the street, giving diversity and articulation to each block with continuous activities distributed along the street (Paumier 2004).

The set of structures in a city center had similarities that connected them, as well as differences that created visual diversity. This diversity was due to the fact that each building was the work of a different designer made for diverse users with their own needs and tastes. Over time, construction methods changed, as did building codes, materials, and economic conditions. However, building blocks in a city center showed visible similarities among the structures, which described the building complex as a cohesive, connected unit. Buildings were generally the same size and with similar materials in similar styles and forms. Although there were differences in the appearance of buildings in most cases, most of the symbols were quite orderly and unique (Francaviglia 1996). As Longstreth (2000) states, these buildings originated from the industrialization of the people and represented the ambition, achievements, and potential of the city of their time. They were a school of competitive imagery that manifested a sense of market competition (Longstreth 2000).

A person walking down the street would quickly walk past different buildings, windows, and entrances, experiencing a great deal of visual variety in a short amount of time. These traditional city centers were pedestrian-oriented in scale because walking was the primary mode of transportation during their development.

The city center was connected to its surroundings and was integrated with the housing. Such centers were considered cities' geographical, socio-economical, and psychological centers (Kunstler 1993).

During the twentieth century, changes in transportation, land use, economy, and population growth had a significant impact on city centers. Although there were complex causal relationships behind these changes, some specific reasons can be stated for the decline of city centers. In the late nineteenth century, the rise of horse-drawn carriages and electric streetcars led to the separation and dispersal of unique places of work, residence, and recreation. In the 1950s, the introduction of cars into the cities facilitated the movement of urban dwellers and business owners. Truck transportation became cheaper and easier, and many commercial activities no longer needed to be located near rivers or train stations. The growth in population and economy accelerated the outward sprawl of the city centers.

An early sign of the decline of city centers was the loss of most of their affluent and original inhabitants. The expansion of the city center pushed profitable economic activities to neighboring units, reducing the quality of the living environment. As the quality of residential houses declined and their lifespan increased, people who could afford them moved out of the city center and started commuting to work daily.

High-rise construction became feasible with the development of contemporary technologies, raising urban land's value. Small-scale residential areas would struggle to survive in such circumstances. As a result, the city center changed from being a vibrant market with a 24-hour activity cycle to one that depends on more specialized, occasional, and regular activities. Land uses like hospitals, courts, offices, and services, as well as legal, specialized, governmental, cultural, and educational organizations and departments, have replaced declining retail and residential uses and the presence of a sizeable number of workers in city centers. As simple daily uses were established and replaced, the city center's life cycle decreased. As people departed the city center, empty street views never returned to how they had been. As a result, the city center's economic context continued to weaken and saw less investment, with just minor commercial applications with little potential for building maintenance.

2.4 Spatial System of City Centers

2.4.1 Single Center Model

The majority of city centers used to be the hub of numerous uses and activities. Typically, the physical form and direction of the growth of the entire city were heavily influenced by the city center. Such a center could be represented in the form of a linear center (e.g., markets in the old cities of Iran) or a centralized space (e.g., important city squares). More importantly, each city is dependent on a single city center. In fact, the single-center model was dominant in most ancient cities and the world, and even today can be seen in some towns. As time passed and communications and interconnections expanded, the relationship between time and space changed. That is, the concentration of resources, facilities, and activities moved from the central core of the cities to the surrounding areas, creating multiple dispersed centers. Later on, sub-city centers were also formed with the movement of people to the suburbs.

2.4.2 Multi-center Model

The transformation of cities in the modern era has resulted in new centers (administrative, cultural, industrial, and entertainment) in addition to the old ones (markets and commercial centers). Gradually, specialized centers got separated from ancient city centers. In many medium and large cities, the multi-center model emerged to meet the specific and differentiated needs of the residents. Communication networks and urban transportation routes play a leading role in creating these centers. In addition, the intersection of roads with each other and with urban squares, subway stations, and transportation terminals pave the way for the concentration of some commercial-service activities. Due to their accessibility and focus, some recreation centers, universities, and other public facilities gradually become city centers around which new city centers emerge. In a multi-center structure or system, specialized centers are developed in each segment and gradually replace the main city center with the same status in terms of quality and quantity (Farid 2010).

2.4.3 Hierarchical Model

In general, city centers follow a specific spatial hierarchy in the context of new urban planning. Claire (as quoted by Saeidnia (2004)) states that the hierarchical model of city centers can be described as follows:

(A) The main city center is considered the city's social, economic, and historic center.

(B) Regional centers (regional shopping centers) are the second primary type of centers for retailing and services after the city's main centers. In addition to the mentioned uses, commercial companies, service organizations, and administrative and cultural institutions are also located in these centers. The population of regional centers ranges from 100 to 500 thousand people and can reach up to one million people in large metropolitan areas. These centers have problems similar to those in the main city centers.

(C) Community centers are placed between regional and neighborhood centers and are of the same size as the town center. They provide monthly goods and services required for the residents. For example, in Tehran, community areas are known by specific names, such as Narmak, Gisha, Gholhak. In contrast, regional areas do not have specific names and are known by their geographical location (e.g., south, north, west, and east).

(D) Neighborhood centers provide commercial-social services at the level of neighborhood units. They often supply daily services and goods required by the residents.

References

Bell D (2005) The emergence of contemporary master plans: Property markets and the value of urban design. J Urban des 10(1):81

Birch EL (2009) Downtown in the new American city. Ann Am Acad Polit Soc Sci 626(1):134–153

Brambilla R (1977) For pedestrians only. Whitney Library of Design

Cowan R (2005) The dictionary of urbanism. Streetwise Press

Farid Y (2010) Geography and urbanism, 8th edn. Tabriz University Publications.

Ford L (2003) America's new downtowns: revitalization or reinvention? Johns Hopkins University Press

Francaviglia RV (1996) Main street revisited: time, space, and image building in small town America. University of Iowa Press

Golany G (1976) New-town planning: principles and practice. Wiley

Hall P (1998) Cities in Civilization. Weidenfeld and Nicolson, Chicago

Jacobs K (2005) The Manchurian main street: are shopping districts inspired by new urbanism a form of cultural brainwashing? Metropolis 24(10):110

Kokabi A (2005) Planning the quality of life in urban centers, definitions and indicators. Master's thesis. Tarbiat Modares University

Kunstler JH (1993) The geography of nowhere: the rise and decline of America's man-made landscape. Simon & Schuster

Langdon P (1994) A better place to live: reshaping the American suburb. University of Massachusetts Press

Library of Congress (1910) Cleveland Federal Building and Post Office [Photograph]. https://www.loc.gov/item/2016800535/

LMN Architects (1998) Urban design policies for city center. Lenexa, Kansas

Longstreth R (2000) The Buildings of main street: A guide to American commercial architecture. Altamira Press, New York

Lund, K. (2013) Chicago street scene [Photograph]. Flickr. https://www.flickr.com/photos/kenlund/11004373093/

Mattbuck4950 (2013) Wembley library [photograph]. Flickr. https://www.flickr.com/photos/36593372@N04/8710422934/

Paumier C (2004) Creating a vibrant city center: urban design and regeneration principles. Urban Land Institute.

Project for Public Spaces (PPS) (2008) Great downtowns fill cities with life. Retrieved from http://www.pps.org/downtowns/

Saeidnia A (2004) The system of urban centers and residential spaces. green book of municipalities guide, vol 4, 3rd edn. Tehran Municipalities Press.

Saidi Rizvani N (2001) Heart of the city, knowledge and involvement. Municipalities J 3(34)

Seattle Parks and Recreation (2014) Splash park [Photograph]. Flickr. https://www.flickr.com/photos/seattleparks/16045790745/. Ghorbanian M, Behzadfar M (2019) Design good streets; based on environmental characteristics. Armanshahr Architecture & Urban Dev J 1(1)

Shakibamanesh A (2008) Promoting city centers quality by using special urban design guidelines. Master's thesis. Iran University of Science and Technology, Tehran

Southworth M (2005) Reinventing main street: from mall to townscape mall. J Urban des 10(2):151–170

Spragg NZ, Bernard (2019) Downtown Toronto. Retrieved October 26, 2022, from https://www.flickr.com/photos/volvob12b/49663498897

Wikimedia Commons (2013) Trajan's market, Rome, Italy. Retrieved 26 26 Oct 26 2022 https://commons.wikimedia.org/wiki/File:Trajan%27s_Market,_Rome,_Italy.jpg

Wikipedia (2008) 29th Street Mall, Boulder, CO [Photograph]. https://en.wikipedia.org/wiki/File:29th_Street_Mall_Boulder_CO.jpg

Wikipedia (2011) Civic center. Retrieved 26 Oct 2022 from https://en.wikipedia.org/wiki/Civic_center

Chapter 3
The Evolution of City Centers

3.1 Introduction

This chapter looks at the development of city centers in Iran and throughout the world. In order to accomplish this, the function and status of city centers during various historical eras are first discussed briefly. The review begins with an examination of the position of city centers in significant Greek and Roman cities. The Sassanid and post-Islamic city centers of Iran are classified and analyzed. As the purpose of this chapter is to evaluate the evolution of city centers to date, various approaches to urban centers and their design in the plans of leading urban planners are reviewed and analyzed. Ebenezer Howard's Garden City, Tony Garnier's Industrial City, Lloyd Wright's Broadacre City, Soria y Mata's Linear City, Pierre Charles L'Enfant growth Centers in Washington, DC, and Le Corbusier, Pierre Jeanneret's joint projects are cited in this regard. In the sections that follow, urban centers after World War II and in the modern era are discussed, as well as new forms of city centers such as suburban centers, shopping centers, indoor shopping malls, fashion centers, etc. Through a careful yet concise examination of the history of changes and developments in city centers during different eras, this chapter attempts to determine the applicability of these lessons to the design and development of effective city centers in the present. This chapter finishes with a consideration of a variety of difficulties in contemporary city centers that are the result of poor design and planning.

3.2 City Centers in Greek and Roman Civilizations

Almost without a doubt, the growth and development of city centers reflect the actual quality of urbanization in many of the world's largest cities throughout history. The Greek agora, the Roman Forum, the medieval business squares, and the Renaissance squares all placed a focus on urban life and provided a stage for the manifestation

A. Shakibamanesh and M. Ghorbanian, *Designing Responsive City Centers*,
Design Science and Innovation, https://doi.org/10.1007/978-981-97-8911-5_3

of history across time. All of these city centers share a similar denominator in their size, vibrancy, and the attributes of their surrounding environment that have made them a "place" (Paumier 2004). Since the discussion of the role and place of city centers in past civilizations requires a comprehensive and complete investigation in its own right, the following sections focus on a brief overview of city centers in the Greek and Roman civilizations.

The primary components of Greek cities were the acropolis, fortifications, agora, residential neighborhoods, one or more recreational and cultural areas, a religious district (if separate from the Acropolis), the pier and port, and the industrial district.

The Acropolis is the first hill or highest point of the city that serves as a defensive location. The term referred to the high center of ancient Greek cities and the fortified castle of many of its colonial settlements. The Acropolis, which once held the entire city, gradually became a holy and religious area of the city (e.g., Athens) or was relocated beyond the municipal limits (as in Miletus).

In Athens, which was the center of Greece's golden age, acropolis was a rocky hill with some of the most magnificent buildings ever constructed. Agora was situated on a gentler slope. It was actually the city center and playground where the Greeks practiced the democracy they had invented. At that time, the Acropolis was the spiritual core of Athens, where the temples of the goddesses (gods) were located, while the Agora was the social, political, and economic center of life (Morris 2013). This split became more distinct following the Greco-Persian Wars (479-490 BC) (see Fig. 3.1).

Fig. 3.1 Agora as a political, social, and commercial center of Greece. *Source* Wikimedia (2015)

Fig. 3.2 Part of the Roman Forum ruins, which once served as the city's beating heart. Authors' Personal Archive, 2022

In Rome, the Forum played the role of the Agora (Fig. 3.2), serving as a center for interaction, discourse, and gathering, as well as for trade and political affairs. The Roman Senate was located around this city center, which later on, after the conversion to Christianity, hosted cathedrals as well (Morris 2013). Romans build the Forums not to emulate the Greeks, but to satisfy the particular requirements of their civic life. However, they were unable to develop a standard shape for Forums as the Greeks accomplished for Agora. In each location, the form of the Forums was independent of previous practices. Notable is the fact that separate Forums relied on the economic systems of their particular cities and, as a result, did not have the same physical form. Some of them were more concerned with business activity, while others were so devoted to administrative formalities that they became a monument collection. Forums were located in the heart of the cities and served as the focal point of everyday life, which took the shape of squares surrounded by columned covered buildings.

3.3 City Centers in Iranian Civilization Before and After Islam

Throughout history, Iranian city centers have had different characteristics. Before Islam, during the Medes, Achaemenid, Parthian, and Sassanian eras, and after Islam, the city center was always regarded as the most vital area, albeit with varying definitions and roles. Since the characteristics of the city center in various historical times are not the subject of this volume, just the location of the city center in the period directly before Islam (the Sassanian era) and the first period following Islam are discussed here.

During the Sassanid empire, the city center was a "government fortress" that became known as Kohandej or Qhandej during Islamic times. The fortress included palaces, the chief fire temple, divans (offices), barracks, reserves, treasury, and food storage facilities. It was built on the highest point of the city and functioned as the city's most important strategic location, protected by enormous walls (Habibi 2014). The castle was a symbol of the Sassanid empire's power and contained a fire temple at its center. In ancient Iran, the city center was more of a ceremonial area serving the government and politics than a social location. However, from a different perspective, smaller centers used by regular people might be identified (rather than aristocrats and courtiers). Such areas were frequently located at the entrances of cities, in front of the gates, or in specific places such as caravansaries, trade venues, sports arenas, ritual ceremonial sites, etc. These small-scale centers in the shape of public squares were frequently constructed at the junction of paths leading to city gates and were utilized for the trading of products. The open areas generated by the junction of these paths with the city gateways typically transformed the square, as the socioeconomic center of the city, into the physical center as well. This period's towns also had religious meeting spaces, which were typically constructed outside the city walls near fire temples because the area in front of the worship sites within the fortress was rarely used by the general populace (Pakzad 2013).

During this time period, the bazaar was not located in the city's center, but rather served as the city's backbone, running from the heart of Sharestan (the middle residential area) to the outer residential area (Rabaz) and forming neighborhoods (Mahalle) along its way. Even though the bazaar was not physically central during this time, it served as the economic and commercial hub of the city through which goods and artifacts were redistributed (Fig. 3.3) (Habibi 2014).

At the beginning of the Islamic era, the central mosque was placed at the center of the city. on that time only settlements with a central mosque were considered cities. Due to the fact that the mosque was the first structure constructed (or converted from a fire temple, a temple, or a church) upon the conquest of a city, it was one of the most important functions and uses of city centers during that time. During this time, the significance of bazaars also doubled. In fact, a central axis as an urban backbone supplanted the central governmental core in pre-Islamic towns, serving as the principal urban element that determined the shape of the city in Islamic cities. This axis influenced the general form of the city so that the placement of other elements was defined in reference to this primary framework. The Square, mosque, school, bathroom, etc. were indispensable parts of the axis. All of these elements were functionally tied to one another, which increased the value of the linear centrality (Fig. 3.4). In the meantime, squares steadily attained a particular position, gaining commercial, cultural, religious, and military applications, among others. In general, the Safavid era can be viewed as the beginning of the flourishing of city squares (as city centers) within the urban structure of Iranian cities (Fig. 3.5).

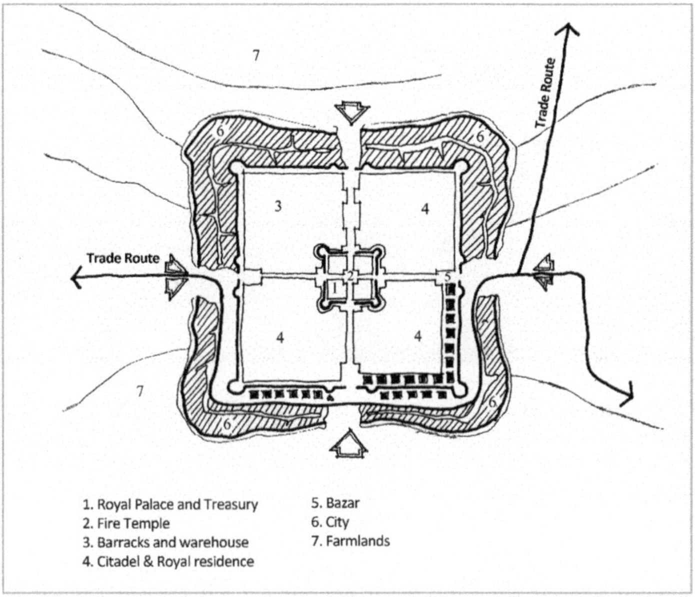

Fig. 3.3 Structure of the Iranian city during the Sassanid era. *Source* Adopted from Habibi (2014)

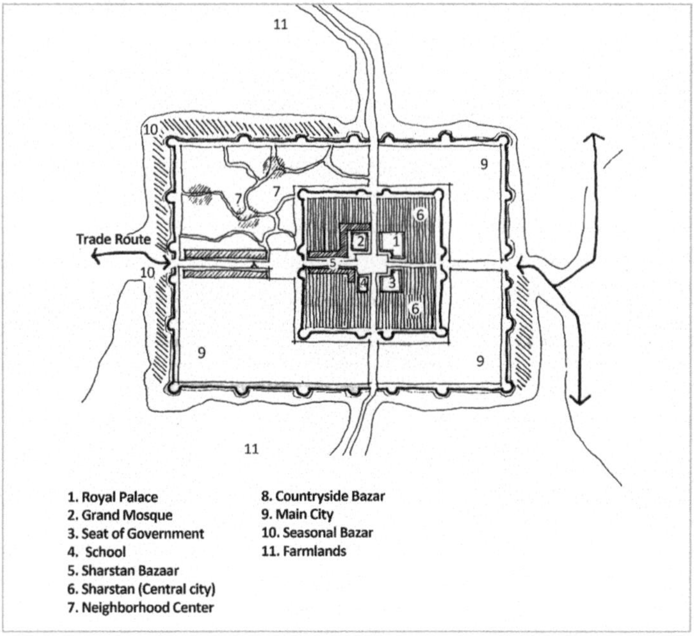

Fig. 3.4 Structure of the Iranian city at the beginning of the Islamic era. *Source* Habibi (2014)

Fig. 3.5 Naghsh-e Jahan Square of Isfahan in 1718 AD, by Bruijn de Cornelis. *Source* Wikimedia Commons (n.d.)

3.4 City Center Plans by Leading Urban Planners

The city center holds a special place in all renowned urban planning initiatives, so that nearly no urban planner has disregarded it. All paths in Ebenezer Howard's Garden City lead to the city center. "Six magnificent boulevards, each measuring 36 m in width, traverse the city from its core to its outskirts and split it into six sections or neighborhoods. A lovely and well-watered garden occupies around 2 hectares in the city's heart, surrounded by the major public buildings, including city halls, concert and music halls, theaters, libraries, museums, painting houses, and hospitals" (Choay 2019) (Fig. 3.6).

In Tony Garnier's industrial city, several main elements are located in the city center. These include administrative services and community halls, exhibitions and theaters, and sports and leisure facilities, as well as city council services, public property registration authority, and courts. Additionally, community halls include a large hall with 3000 seats and smaller side halls. All of these halls open onto a big porch, which forms an indoor promenade in the heart of the city where a great number of people can congregate regardless of weather conditions (Choay 2019).

Even Frank Lloyd Wright's Broadacre City, which lacks the density and cohesion of conventional cities, includes the city center idea. In this plan, the city center is spread over a large area and includes a 50-story office building, a park, sports fields, a zoo, and an aquarium.

Le Corbusier's urban planning ideas are the opposite of Frank Lloyd Wright, as they encourage urban density and recommend towers instead of human-scale spaces that are enjoyable to walk through. In a proposal given by Le Corbusier and Pierre Jeanneret in 1922, the city center comprised 24 towers, each with 60 stories and arms measuring 175 m. Each of these enormous buildings could accommodate 10,000 to 50,000 employees. Thus, the population density in the city core reached 30,000 per hectare.

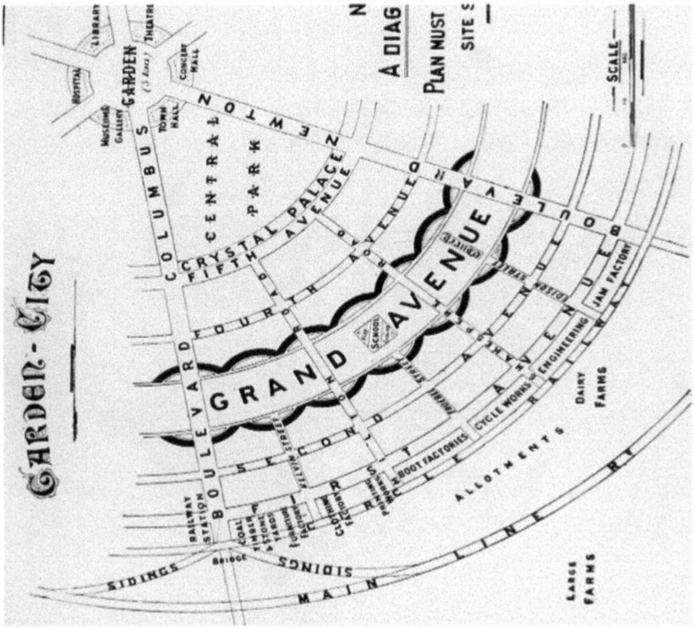

Fig. 3.6 City center in Howard's Garden City. *Source* Internet archive book images (n.d)

In Arturo Soria y Mata's linear cities, however, instead of high-density traditional cities that are built around a central nucleus, a completely different type of center is proposed, consisting of a narrow strip of unlimited length with one or more railways going through its axis. According to him, a perfect city is dispersed along a single road and is around 500 m wide. The main street must be at least 40 m wide and planted (Benevolo 1977). In fact, he substitutes the central street for the center core.

As can be seen, the city center has been one of the primary elements urban planners have viewed as the gravitational center of the city, which can play an accelerating function. According to this viewpoint, if this location is chosen wisely and is powerful enough, it can have an impact on the surrounding areas and alter the entire city. This strategy was utilized by Pierre Charles L'Enfant when designing the layout for Washington. He isolated the White House and the Capitol Building from the western region of the city (Georgetown) on purpose so that each would serve as an incentive for the development of a broader territory (Fig. 3.7) (Lynch 1984). The same concept applies to the "Growth Pole" theory. Thus, for the economic revitalization of a region, several investments are made to reinforce a certain location, so that growth will spread from that point to other parts of the region for the objectives of development (Saidi Rizvani 2001).

Fig. 3.7 L'Enfant's plan for the separation and optimal arrangement of two key growth poles (White House Building and Capitol Building in Washington, DC). *Source* Yang (2012)

3.5 City Centers in the Contemporary Age

Most of the core areas of major European cities, which are reminiscent of the Middle Ages or the Renaissance, are now symbolic of a combination of historical urban planning practices and contemporary urban planning. Typically, the streets and boulevards of this section of the city are constructed on the old city walls and encompass the old city gates (Fig. 3.8). Typically, a number of historical landmarks can be found inside these city centers.

Medieval churches and cathedrals, formerly the highest urban buildings, contribute to the magnificence of the central parts of European cities. Today, these churches and cathedral in the center of European cities serve as tourist attractions. The Notre Dame Cathedral in Paris, San Marco in Venice, and Milan Cathedral (Duomo of Milan) are some examples (Fig. 3.9).

The historic palaces of kings are another attraction in the urban cores of European cities. The Louvre in Paris, Castello in Milan, and the Kremlin in Moscow are some examples. Throughout history, the historical centers of European cities were bustling marketplaces, with the busiest ones forming in front of churches. In many European cities, market days are held in the same location as they were for instance 500 years ago, where in some cases today's huge enterprises and wealthy merchants conduct

Fig. 3.8 By a historic gate, medieval Siena is divided from the new city. The picture on the right was captured from within the medieval city, while the image on the left was captured from outside. Siena, Italy. Authors' Personal Archive, 2019

Fig. 3.9 The Duomo Cathedral in center of historic Milan tells a story of faith and art spanning over six centuries. Authors' Personal Archive, 2019

business. In Zürich, for instance, the Saturday market sets up adjacent to the Main River (Fig. 3.10) (Saidi Rizvani 2001).

Since the Middle Ages, the urban cores of European cities have traditionally served multiple urban functions simultaneously. The central part of the city is the main center for retail uses, residences of various social classes, and handicraft shops. As a result, it is rather difficult to define and separate residential, industrial, and commercial districts in European cities, as these urban roles are extensively intertwined (Shakoi 2011).

From the earliest days of the birth and formation of American cities, distinct city centers with comparable characteristics emerged. In the 19th and early twentieth

Fig. 3.10 Zürich's best flea market is Kanzlei. Saturday mornings find vendors setting up their stalls. At this flea market, local residents can not only purchase treasures, but also resell them. Here you will find everything imaginable, from fur coats to cooking items, electronic equipment to leather bags and jewelry. Authors' Personal Archive, 2022

centuries, as the United States was in the process of urbanization, a typical kind of city center became popular (Francaviglia 1996; Ford 2003). The first skyscrapers were constructed in Chicago's central business district (the Loop) in 1872. In American cities, the central business districts (CBDs) were the focal point of commercial and service activities and were generally devoid of residential uses. The development of the urban core in these cities was tied to the growth of the tertiary sector of the economy (services) (Farid 2010). The stronger the sector, the greater the demand for city center housing. This strategy put pressure on city centers, resulting in their progressive decline. Since 1960, American urban planners have focused on rebuilding city centers. Since then, new urban planning theories have been focused on three primary axes (Fig. 3.11).

(A) Strengthening social and cultural standards and values in urban planning and design;
(B) Strengthening and developing urban spaces and public leisure facilities; and
(C) Adjusting car movement and expanding pedestrian spaces in city centers.

This movement also affected European urban management, forcing them to take action to preserve their valuable city centers. It was not until around 1975 that almost all major European cities banned cars from entering much of their historic and central areas. This trend was accompanied by the expansion of planning and organizing pedestrian spaces (Saeidnia 2004) (Fig. 3.12)

Studies show that in the 1970s and 1980s, many major cities in Europe and the United States shifted their approach to city centers toward social, economic, and quality-centered goals and values. This change in the approach and procedure in design and planning has necessarily been accompanied by the expansion and

Fig. 3.11 New York City Times Square: an example of an American city center. *Source* Terabass (2012)

organization of pedestrian spaces, public open spaces, etc., and has been carried out to achieve various economic, environmental, social, and qualitative goals.

Over the past three decades, several strategies for organizing city centers have been applied in many cities across the globe. Rotterdam is the first municipality to take advantage of the opportunity to revitalize the city by removing vehicles from city centers (Garvin 2002). In the 1970s, mixed-use development, pedestrian facilities, and the growth of public transportation were prioritized in the Central Core Area of downtown Toronto, Canada, as seen by the emphasis on mixed-use development. During the 1970s, efforts were established to renovate the city core of Stockholm, thereby enhancing the quality of life and urban appeal.

Since 1981, a plan for the restoration of the worn-out and polluted center of Athens has been implemented based on the expansion of pedestrian zones and the improvement of the city's physical and social appearance. In 1983, high-rise buildings in the San Francisco city center were restricted by popular vote. Other prominent examples of successful and attractive pedestrian spaces in city centers include Portland, USA, which hosts a variety of ethnic events and markets, attracting many people who may want to participate (Saeidnia 2004). In developing countries, on the other hand, due to the economic profitability of city center activities, urban cores are becoming increasingly congested as the incompatibility between activities and available space increases. The activities of the informal sector of the economy creep into the city centers, resulting in the decline of the quality of these areas. However, certain cities

Fig. 3.12 In some historical areas of Milan, only taxis, trams, and electric vehicles are permitted (above). However, other areas surrounding the historic core of the city are designated for pedestrian movement (bottom). In all of these streets, emergency movements (for police, firefighting, and emergency vehicles) are considered. Authors' Personal Archive, 2019

in these nations have made valuable and occasionally fruitful steps to revitalize the historical contexts, which are typically located in the city's core.

3.5.1 The Formation of Suburban Centers

Transportation technology advancements in the nineteenth century led to the expansion of cities beyond walking distance. Similarly, the introduction of railroads and urban trams encouraged the growth of suburban towns outside of cities. After World War II in the late 1940s and into the 1950s, low-density car-based communities, including single-family homes, became commonplace in new residential areas in the United States. The shift in focus after World War II was not simply a shift to lower densities but was part of a rise to the modernist style of architecture and planning. Modernism in architecture grew popular as it shifted toward practical aesthetics and emphasized the separation of spaces throughout the planning stage. With the introduction of the new urban planning concept, the spatial (three-dimensional) interaction between streets and buildings underwent a radical transformation. The configuration of streets and buildings was no longer perceived as a network. Instead, they were divided and the roadways followed their fluid geometry, resulting in three-dimensional buildings within the spaces (Southworth 2005).

The historic patterns of city center design with their compact blocks and close-knit structures inside a grid of vertical angles have become obsolete. The new projects incorporate major highways and arterials. Large, expansive lots and Auto-Oriented Infrastructure were characteristics of modernism (Southworth and Owens 1993). Dense city centers and main streets are marginalized in modernism, as overtly stated by the famous modernist architect, Le Corbusier, who recommended a city without streets (Marshall 2004).

During this period, cars and suburbanization were viewed as elements speeding the demise of city centers (Longstreth 2000). The commercial activity was transferred along highway corridors. Later, it was recognized that this overreliance on highways undermined the cohesiveness of urban society (Duany et al. 2001) (Fig. 3.13)

Throughout history, the shape of city centers has been strongly influenced by the modes of transportation employed during development. Due to a change from pedestrian-driven models to reliance on personal motor cars, urban congestion has decreased. This development was mostly the result of the demand for wider places for driving and parking vehicles. In general, during this period, the need for the city center to create a sense of place and meet the needs of suburban residents was still felt in the suburbs (Bohl and Schwanke 2002). In the subsequent sections, we will discuss various approaches to this demand.

Fig. 3.13 Plymouth and Northville, examples of suburban centers in Detroit, US. *Source* Hudkina (2012)

(A) **Shopping Centers**

An important feature of new suburbs is the separation of uses from each other. Different functions were considered for different areas through zoning, in such a way that the combination and mixing of land uses rarely occurred in a place (Duany et al. 2001). As mixed uses disappear and got dispersed over large areas, in most cases there was no clear central business district. In the absence of the city center, major shopping areas were identified as urban community centers (Scholl and Williams 2005). These shopping centers were largely different from their predecessors in terms of shape, personality, and development pattern (Duany et al 2001). One of the most obvious differences was that modern shopping centers were to provide services to buyers who came there by car. Therefore, there were little or no pedestrian connections with adjacent areas. Sidewalks were no longer used to connect centers to neighborhoods and residential areas, even if there were residents living nearby.

Such centers were placed along main roads and usually at the intersection of two busy roads. Large parking lots were constructed in front of buildings, giving buyers enough space to park their cars, and keeping the buildings away from the street edge (Duany et al. 2001; Kunstler 1993). The location of commercial buildings near parking lots changed the definition of the street. The shops were no longer facing each other along the streets of the business district, and the central part of the city was not made of corridors that could be crossed when shopping. Instead, shoppers entered a large, open area, with all the shops on one side facing them. Shopping malls were located close to the roads and car traffic passed by (not through) the city center. City centers no longer followed urban block patterns, with no traditional urban context being seen there (Southworth 2005).

The architecture of shopping center buildings was entirely distinct from that of earlier urban structures. Instead of blocks composed of many small buildings, they were structurally distinct and separated by walls. A shopping center was comprised of one or a limited number of large buildings, which were vast structures with enormous ground floors that could be utilized for a variety of commercial purposes. They were easily divisible into smaller areas for a number of commercial uses. The typical structure of shopping center buildings consisted of a concrete floor and a flat ceiling on the load-bearing walls on the sides and the rear, as well as columns in the front and central sections. Buildings were usually one-story and accommodated no use other than buying and selling (Duany et al. 2001; Francaviglia 1996).

Instead of gradually evolving over time, as traditional city centers did, shopping centers were built by one person at a time. The design of each center was completely uniform and prepared by a single architect. The main motivation for building a shopping center was profitability, and the result was buildings that were not as principled as the traditional city center buildings and in some cases were considered disposable structures (Kunstler 1993). However, their exteriors were carefully designed to have an attractive image. More expensive and good-looking materials were generally used in this part of the building. In some cases, the façade of the main building was the same for all floors, creating a monotonous and dull appearance (Fig. 3.14).

Fig. 3.14 Examples of shopping center in the US. *Source* Cinema Treasures (n.d.)

In terms of character, a shopping center was mostly known for the concepts of a particular style. The objective was to provide a sense of place to the center (Southworth 2005). In truth, it was an instance of separate commercial activity cooperating and agreeing to abide by general rules to promote their common prosperity (Gruen and Smith 1960).

After World War II, private commercial complexes in many suburban areas sought to establish community centers because no location has the required physical attributes to serve as a central location. In most cases, this was done unplanned with no intention on the part of local governments. Thus, in addition to shopping complexes, enclosed shopping malls were developed primarily to serve the surrounding residential communities (Bohl & Schwanke 2002).

(B) **Enclosed Shopping Malls**

Over time, shopping center development patterns also expanded, and in the 1960s, enclosed shopping malls of greater size emerged. As opposed to shopping centers, which were oriented outward, they were centralized (toward parking spaces). Most shop fronts were directed to an inner sidewalk, which was usually closed to cars. Shops were placed opposite each other along the enclosed space, as in the street in a traditional center. The width of the central bazaar resembled a street, except that it was closed to cars and was only used by pedestrians. They generally had a certain kind of layout known as the Dumbbell Plan, where major shopping malls and hubs were placed at both ends of the open space (Fig. 3.15) (Booth et al. 2001).

These malls were built in accordance with a Merchandising Plan, in which anchor stores were situated such that shoppers had to go past smaller stores to reach them. The designers attempted to make the storefronts visible to the greatest number of

Fig. 3.15 The arcade-enclosed shopping area in downtown Cleveland. *Source* NARA (n.d.)

passing customers, ensuring that the distance between the main shops did not exceed 182 m. In order to prevent their uniformity, plazas and courtyards were sometimes incorporated into the design of these centers. In early malls, pedestrian spaces had a width of less than 15 m, with a minimum of 3 m in front of shops on each side, and a minimum of 3 m for a central green space, seats, and other amenities. In later malls, the distance increased to 15 m, which was a substantial increase (Maitland 1985). Such centers were a clear reflection of the creation of a linear axis as the city core. The parking lots were often situated all around the center. As the buildings were internally orientated, the loading areas and outside walls were facing the parking lots. The enclosed malls have little in common with the surrounding streets. At the time, these malls were the most common form of local shopping centers and the most frequent shopping destinations (Southworth 2005).

The architectural style of private retail malls built after World War II was notably distinct from that of conventional city centers. Their uses, ownership, and public places were the primary differences. All land within a modern retail mall, including roads, open spaces, and sidewalks, is privately owned. Every site designated for public use, such as plazas and parks, is actually owned and managed by a private commercial entity (Blum 2005). Under these conditions of ownership, the managers have access to and authority over the activities occurring within the place. Any formal or informal group event or ceremony requires the approval of the property owners.

In private shopping malls, the developers and owners deem their facilities just for shopping without taking adequate measures to encourage people to stay there (Southworth 2005). Victor Gruen, one of the pioneers of mall design, believes that malls are more than just purchase and sale places, but are "shopping towns" that must incorporate a combination of mixed uses (social, civic, cultural, etc.). To Gruen, enclosed shopping malls can be the manifestation of suburban social life, an alternative to the city center, and a factor that can prevent sprawl. However, most malls rarely have uses other than shopping, eating, and having fun. Shopping centers and micro-malls are private, and owners do not have to create a public space in such malls (Gruen & Smith 1960; Smiley 2002).

In "The Great Good Places", Ray Oldenburg points to reasons that have prevented enclosed shopping malls from acting as community and gathering centers. In addition to private ownership, lack of civic uses, as well as encouraging buyers not to stay or stop, such reasons include the size of the malls as an influential factor. For a place to lead to community development, it should create meeting opportunities for people in a small community who usually know each other. Given their large scale, suburban enclosed shopping malls usually require high sales, which entail a large regional market. Also, the items sold in the mall are usually bought in lower quantities compared to items purchased on a daily basis. Creating a sense of collective and social interaction in the mall is unlikely because it is local and is based on a large number of occasional customers (Oldenburg 1997).

The average lifespan of an enclosed shopping mall in the United States is currently less than ten years and the number of closed markets is predicted to increase across the country. In 2002, Urban Land Institute estimated that 15–20% of these centers would have to close in the near future. Current patterns show that many more of these centers need to be revised and developed based on modern ideas (Schwanke 2003).

(C) **Emergence of Centers with New Approaches**

New types of shopping centers (e.g., regional centers, super-regional centers, fashion centers, and outlet centers) were created after shopping malls became obsolete from the standpoint of the sales and development industry and trade. One of the most famous of these centers in the late 1990s was a lifestyle center as a luxurious place aimed at attracting people from an extensive area. Like early strip centers, it includes a row of shop facades along the outer sidewalks facing a large car park. However, the views of these shops are more intricate and costly than those of conventional shopping malls, and they vary from shop to shop. This kind of centralized organization emerges with the aim of making shops more visible and responding to the evolving lifestyles of contemporary suburban residents by providing quick access to shops. Lifestyle centers also include buildings for recreational and dining purposes (Weber and Palmer 2005).

Another new concept in the design of shopping centers originated from lifestyle centers. These developments, known as "town centers" were based on the concept of creating shopping malls similar to a traditional American small-town downtown (Southworth 2005). The first shopping mall built on this concept emerged in the 1970s.

James Rouse was one of the pioneers in this type of development, which was later called "Festival Marketplaces." Though the style was initially successful, turned out to be a failure. The difference between town centers and previous city center developments was that cars were not allowed on its streets, a factor which proved to be an important factor in the failure of such centers. Vehicle access to the streets of city centers plays a vital role in their success (Smiley 2002; Garvin 2002).

Recent town centers use some concepts of festival marketplaces as well as shopping malls. They are usually large scale and include more extensive areas compared to festival marketplaces, and have wider streets with the ability to respond to car and traffic and parking needs on the street, as in enclosed shopping malls. They have dumbbell-shaped plans with attractive uses such as shopping centers, cinemas, and some facilities located in the surrounding streets (Bohl & Schwanke 2002).

Shopping malls in town centers, while in harmony with the overall structure of the city, are designed in such a way that allows components, elements, and details to be formed in them, increasing their resemblance to a real center of a town (Scholl and Williams 2005). Spaces such as town squares are included in most designs using traditional street decorations and architectural styles. Sometimes the facades of buildings are different for most shops so that the streets seem to consist of many small buildings as in traditional small towns, even if they are not (Bohl and Schwanke 2002). The best opportunities for densely populated suburban centers are found in existing business hubs. These places provide a chance to bring together a mix of uses (Booth et al. 2001).

Many of the existing suburban shopping areas are built around the boundaries of older shopping malls, with other uses centered around them. Many of these malls have been closed or are struggling to maintain themselves, and as stated in Congress for the New Urbanism, have created good opportunities for redevelopment and infill development (Schwanke 2003).

3.6 Problems and Issues in City Centers

The primary concerns and problems of city centers must first be addressed from a structural and functional point of view in order to accomplish the demands, goals, and appropriate organization techniques. The most important issues in this area are discussed below.

3.6.1 Conflicts with the General Structure of Cities

City centers in the old urban planning model were mainly based on human scales, pedestrian movement, craftsmanship economics, and traditional materials and architecture, while newer developments are based on industrial production and the expansion of the car traffic network. This situation has created a kind of structural heterogeneity in the surrounding contexts. Resolving this contradiction and conflict is one of the problems and dilemmas of new urban planning in large and old cities alike.

3.6.2 Density and Concentration of Activities

Accumulation of factors attracting capital and activity in the city center, as well as scarcity and complete occupation of lands, have led to the high concentration of a variety of activities in the area. This has led to excessive land consumption and fragmentation, the complexity of the land ownership and rent system, as well as the inappropriate and extreme use of space (and the resulting inconsistency of spaces and activities) which can lead to other conflicts and inconsistencies (Saeidnia 2004) (Fig. 3.16).

3.6.3 Loss of Nighttime and Daytime Population Balance

The lack of balance in the daytime and nighttime population balance due to the structure of activities governing the city center leads to the creation of defenseless and non-dynamic spaces and ultimately the weakening of social security, vitality, spatial diversity, and so on. This is mostly due to the incorrect location of activities and lack of proper distribution of functions with due regard to the active time of each. As a result of the simultaneous closure of a number of activities and the resulting eviction of the population, the collective vitality and urban life of the city center during night hours fade away.

3.6.4 Economic Problems

Observations show that the large capital owners and advanced production units are inclined to move from the central area to new commercial and service areas, and

Fig. 3.16 Density and concentration of activities in the central part of the city, Tabriz, Iran. Authors' Personal Archive, 2016

conversely, owners of small capitals and traditional and sometimes abandoned ancillary activities are eager to stay. Retail attractions as a factor that can increase social mobility in the region are declining, while unproductive and insignificant occupations are attracted to the region. From this point of view, the evolution of production and commercial activities in terms of the type of activity, amount of capital and return, production method and the like can be good subjects for further research (Mehdizadeh 2016).

3.6.5 Physical Decline

Some available contexts, equipment, and facilities have physically deteriorated and even been destroyed due to the age of the region and the indifference of residents and property owners toward its maintenance. This circumstance could result in death and the destruction of the context if nothing is done to avoid it. It is also capable of causing physical corruption to spread to nearby areas (Fig. 3.17).

3.6.6 Reduction of Neighborhood and Residential Values

Several city centers that were formerly regarded as the aristocratic and prestigious neighborhoods of the city have steadily degenerated into worthless and destitute places. Many owners and traders in the area have moved to other urban areas, despite maintaining their local employment. This separation of residential and intrinsic values from false commercial values has resulted in the destruction of place values.

Fig. 3.17 Physical exhaustion in the city center, Tabriz. Authors' Personal Archive, 2017

3.6.7 Social Replacement

A combination of unfavorable environmental and economic factors has caused unhealthy, profiteering, and parasitic individuals, such as moneylenders, usurers, displaced people, unemployed immigrants, and the homeless, as well as numerous unproductive and hidden activities, such as hoarding, speculation, smuggling, production of counterfeit goods, black market, etc., to concentrate in the central areas in the form of networks. The existence and prevalence of such phenomena give the area a mysterious, negative and, of course, powerful character, which confronts any kind of urban planning and organization with different kinds of unconventional and unknown problems and resistances (Mehdizadeh 2016). Social displacement in the sense of the departure of the high-income resident population with a sense of belonging and their replacement by the population without a sense of place has, in reality, contributed to the physical and social collapse of the city center context.

3.6.8 Environmental Conditions

The center regions of the city have collected the greatest and most severe forms of environmental contamination, according to the available evidence. Many small industrial activities and workshops are located in this area, which is incompatible with the surrounding environment and difficult to relocate and rebuild.

3.6.9 Traffic Concerns and Space and Area Limitations

It is obvious that city centers have limited areas. Therefore, over time, and in parallel with the growth and development of the city and the population, the pressure on this area increases because of reduced space on the one hand and increased demand for space on the other and finally increased daily trips to such areas. This situation has worsened especially after the arrival of cars that has brought about the need for parking spaces. It is worth noting that according to studies conducted in different countries, on average, each personal vehicle is used only 400 h per year and is parked 95% of the time of the year (at the origin or destination). What to consider is that, in general, most owners of private cars tend to drive to the central areas of cities. Lack of facilities for pedestrian traffic and the weakness of public transportation systems are other problems of urban centers in this area.

3.6.10 Cultural Issues and Concentration of Significant and Symbolic Urban Elements

The central part of the city is one of the main factors and manifestations of historical and artistic values and is the embodiment of symbols, signs, and identity-related memories for the whole city and its inhabitants. These identity-creating elements that are located at the edge of the main axes and squares of these contexts limit the manipulation of networks in these areas. Therefore, solving the problems of these textures logically and consciously requires committed experts who are well-informed about such elements (Fig. 3.18).

In urban planning, cultural and visual values, local affiliations, and public valuation criteria are referred to as "quality indicators" while, communication or economic functions, densities, and the like are called "quantitative indicators." Therefore, quantitative indicators are not enough to determine the city center boundaries, and reference to qualitative indicators is also needed.

Other issues and problems related to the design of city centers include the following ones (Tavasoli 1997; Bahreini 2012).

- Incompatibility of spaces and activities (such as the establishment of workshops in residential units in the city center);
- Incompatibility of forms and functions;
- Lack of coherent and integrated forms in city centers;
- Extension of production, industrial, commercial, and service activities into public spaces;
- Lack of space for development and difficulty of design in the existing compact contexts;
- Lack of pedestrian-oriented pathways in city centers;
- Issues related to the ownership of plots (multiplicity of owners, the small size of plots, lack of effective regulations and strategies for dealing with abandoned lands, ruined structures, etc.);

Fig. 3.18 Identity-related and cultural value in need of preservation in city centers; Gorjiha Bazaar (Gorjilar) in Tabriz (Right); Hassan Abad Square, Tehran (Left). Authors' Personal Archive, 2017

- Expensiveness of land and buildings, which makes it difficult to implement urban restoration and development plans;
- Social insecurity, especially in public open spaces of city centers, parking lots, etc.;
- Low safety due to low quality of the flooring, furniture, and lighting, and lack of proper separation of pedestrian and car territories, etc.;
- High building density and hence lack of human scale in city centers;
- Low-quality facades;
- Low width of sidewalks and their limitation due to improper placement of advertising posts, stalls, utility installations, etc.;
- Lack of spaces for people to stop and interact on the sidewalk, whether formal (benches) or informal (stairs, edges, etc.).

The occurrence of these issues and other problems resulting from the introduction of modernist thought into city centers compelled planners to focus on the issue of riding access to these centers and the adaptation of the city center context accordingly without regard for social, cultural, or historical values. After the failures of this way of thinking in urban planning, city centers must now be viewed as the most important or at least one of the most important elements and components of a city, with social, economic, environmental, and cultural dimensions considered alongside physical and functional design factors. In addition to their primary function as the hub of activities and uses, city centers are the arena of social interactions and the cultural-historical identity of its citizens. Therefore, they are closely tied to recreational areas and tourism-related activities. This crucial objective demands urban designers to pay close attention to a variety of urban living quality standards in the city center.

References

Bahreini SH (2012) Analysis of urban spaces: In relation to the behavioral patterns of users and criteria for design, 9th edn. Tehran University Press, Tehran, Iran

Benevolo L (1977) History of modern architecture. MIT Press

Blum A (2005) The mall goes undercover: it now looks like a city street. Slate. Retrieved from http://www.slate.com/id/2116246

Bohl CC, Schwanke D (2002) Place making: developing town centers, main streets, and urban villages. Urban Land Institute.

Booth G, Urban Land Institute (2001) Transforming suburban business districts. Urban Land Institute

Choay F (2019) Urbanisme, utopies et réalités-Une anthologie. Média Diffusion

Duany, A., Plater-Zyberk, E., & Speck, J. (2001). Suburban nation: The rise of sprawl and the decline of the American dream. North Point Press.

Farid Y (2010. Geography and urbanism, 8th edn. Tabriz University Publications

Ford L (2003) America's new downtowns: Revitalization or reinvention? Johns Hopkins University Press

Francaviglia RV (1996) Main street revisited: time, space, and image building in small town America. University of Iowa Press

Garvin A (2002) The American City: What Works, What Doesn't (2nd ed.). McGraw-Hill.

Gruen, V., & Smith, L. (1960). Shopping Towns USA: The Planning of Shopping Centers. Reinhold Pub. Corp.

Habibi SM (2014) Az shar ta shahr; a historical analysis of the concept of the city and its physical appearance: Thinking and impression. Tehran University Press

Hudkina J (2012) Detroit's suburban centers: Plymouth & Northville. SkyscraperPage Forums. Retrieved from http://forum.skyscraperpage.com/showthread.php?t=170493

Kunstler JH (1993) The geography of nowhere: The rise and decline of America's man-made landscape. Simon & Schuster

Longstreth R (2000) The buildings of main street: a guide to American commercial architecture. Altamira Press, New York

Lynch K (1984) Good city form. MIT Press

Maitland B (1985) Shopping malls: planning and design. Nichols, New York

Marshall S (2004) Streets and patterns: The structure of urban geometry. Spon, New York

Mehdizadeh J (2016) Strategic planning of urban development (recent global experiences and its place in Iran). Architecture and Urban Planning Studies and Research Center of Iran, Iran

Morris AEJ (2013) History of urban form: before the industrial revolution. Routledge

Oldenburg R (1997) The great good place: cafés, coffee shops, community centers, beauty parlors, general stores, bars, hangouts, and how they get you through the day, 2nd edn. Marlowe & Co.

Pakzad J (2013) Urban space design guidelines in Iran. Ministry of Housing and Urban Development Press

Paumier C (2004) Creating a vibrant city center: urban design and regeneration principles. Urban Land Institute

Saeidnia A (2004) The system of urban centers and residential spaces. In: Green book of municipalities guide, vol 4, 3rd edn. Tehran Municipalities Press.

Saidi Rizvani N (2001) Heart of the city, knowledge and involvement. Municipalities J 3(34)

Scholl DC, Williams RB (2005) A choice of lifestyles: Though the mall is here to stay, lifestyle retail centers are catering to consumers who want a complete 'experience.' Urban Land 64(10):88–93

Schwanke D (2003) Mixed-use development handbook. Urban Land Inst

Shakoi H (2011) New perspectives in urban geography. Samt Press, Tehran

Smiley DJ (2002) Sprawl and public space: redressing the mall, 1st edn. National Endowment for the Arts

Southworth M (2005) Reinventing main street: From mall to townscape mall. J Urban des 10(2):151–170

Southworth M, Owens PM (1993) The evolving metropolis: studies of community, neighborhood, and street form at the urban edge. J Am Plann Assoc 59(3):271

Tavasoli M (1997) Urban design in the central part of Tehran. Iran Urban Planning and Architecture Studies and Research Center Press, Tehran

Terabass (2012) Times square, New York [Photograph]. Wikipedia. https://en.wikipedia.org/wiki/File:New_york_times_square-terabass.jpg

Weber J, Palmer AT (2005) How the net is remaking the mall: to lure web-savvy shoppers, retailers are turning to lifestyle centers. Business Week. Retrieved 7 June 2008 from http://www.businessweek.com/magazine/content/05_19/b3932105_mz057.htm

Wikimedia, Commons (2015) The story of the greatest nations, from the dawn of history to the twentieth century: a comprehensive history, founded upon the leading authorities, including a complete chronology of the world, from: https://commons.wikimedia.org/wiki/File:The_story_of_the_greatest_nations,_from_the_dawn_of_history_to_the_twentieth_century_-_a_comprehensive_history,_founded_upon_the_leading_authorities,_including_a_complete_chronology_of_the_world,_and_%2814578689288%29.jpg

Yang J (2012) Building America from scratch. The National Endowment for the Humanities. Retrieved 25 Aug 2022, from http://www.neh.gov/divisions/research/featured-project/building-america-scratch-summer-work-thomas-jefferson

Chapter 4
Manifestation of the Spirit of City Centers on Main Streets

4.1 Introduction

Streets are the current rivers of cities, where we gather and go through toward the city center. This simplest element of urban form comprises the most complicated system of human life and is represented in the minds of inhabitants as a dense hub of diverse activities and life opportunities. The main street is the heart of the community and has its own history and character. It is a location that the community perceives as the city's center, and as such, it must meet the requirements of an ideal and desirable city center. Today, however, many main streets lack these characteristics and, along with other urban environments, have lost their vitality and dynamism.

This chapter provides a brief summary of the history of streets from ancient Greece (Hellenistic cities) to the Industrial Revolution and contemporary eras and discusses the qualities of these rich scenes of collective life. After introducing the significance of streets as simple urban forms, the next sections explain their social, functional, and aesthetic intricacies by referencing urban studies specialists. Then, the conceptual position and general quality of main streets in the past and present, which have followed a declining trend, are explained. Saving and revitalizing these common identities of civil society demands an understanding of their history and a strategy that can lead to the strengthening of various patterns of social interaction and vibrancy on the main streets. Thus, these valuable urban spaces will be able to promote social activity and aesthetic experiences. In this regard, various concepts, such as new urbanism, urban renewal, regionalism, and the preservation movement, are introduced. The theoretical explanation of the specified ways engages the reader's thoughts with the advantages and disadvantages of each as well as the possibility of utilizing their benefits.

A. Shakibamanesh and M. Ghorbanian, *Designing Responsive City Centers*,
Design Science and Innovation, https://doi.org/10.1007/978-981-97-8911-5_4

4.2 An Introduction to the History of Streets

4.2.1 From Greece to Rome (Hellenistic Cities)

During the Hellenistic Period, urban spaces, especially in cities such as Athens were increasingly equipped with more luxurious and magnificent buildings, and the width of the streets, which had not exceeded 4 m, increased to 6 m. During this time, the width of some main streets, such as the Canopy in Alexandria, reached 32 m. The kings of the Hellenistic era displayed their wealth in urban buildings. Moreover, the wealthy also spent considerable sums of money to decorate their city to become famous as charitable persons. The streets were richly decorated, mostly with a row of columns. The buildings on both sides were also built according to certain criteria and with certain heights and shapes. The city was adorned with trees, and the edges of balconies and balconies were ornamented with greenery. The use of benches and equipment in open spaces and the construction and installation of inscriptions, stone writings, and statues of the living or the dead also spread widely throughout the streets.

In Rome, like Athens, there was a tendency to build huge buildings. Luxurious theaters, promenades, markets, libraries, baths, and numerous passages could be seen next to the government headquarters. About 450 BC, when Roman laws were inscribed on the Twelve Tables, regulations were enacted for the construction of urban thoroughfares. The width of the main passages was set at a maximum of 2.90 m, and the houses could have balconies on the upper floors (Shakibamanesh 2008). Defining different types of roads and passages became common since then. For example, while "Actus" had the capacity for only one chariot to pass, the "Appian Way (Via Appia)" was considered for the passage of two chariots next to each other (Fig. 4.1). Later in the empire, the relative width of the passages increased, so that the width of the main passages reached 4 to 8 m. In the eastern part of the empire, some cities also had arched (indoor) malls. Some historians trace indoor malls in the city of Antioch to 360 AD with lighting provided by a public lighting system at night extending up to 25 km. In the city of Rome, however, the passages were not very wide, averaging 6 to 7 m, and in a few cases up to 8 m. In Rome, the streets of the city center often ended in "Forum."

During the empire, the streets of Rome were paved with polygonal flint stones, while the squares had travertine paving. Wider streets had sidewalks. There were heavy poles in these streets and squares that often had inscriptions and were either erected in honor of the Gods and kings or used to indicate the name of the areas and the places in which they were located. In some cases, they were used as signs and showed the distance. In many of Rome's streets, fountains (the most prominent sign of the "prince's" power) could be seen. These lavish works, in which the arts of sculpture and inscription were used together, showed the emperor's generosity and were built to praise the services the emperor rendered to his city. Full-length sculptures or on horseback were also landmarks in the decoration of streets and

Fig. 4.1 The Appian Way (via Appia) is one of the earliest and strategically most important Roman roads of the ancient republic. It connected Rome to Brindisi, in southeast Italy, Rome, Italy. *Source* Authors' Personal Archive, 2019

squares. The use of objects and precious materials such as gold, ivory, and colored marble in the decoration of streets of this period was favored by kings and officials.

4.2.2 Middle Ages

During Middle Ages, other essential urban equipment, such as fountains, stone fences, and religious infrastructure, that were considered essential, could be seen on streets, while other urban furniture and installations were rare. Also, all artistic manifestations in the cities of this period were created and demonstrated with a God-centered attitude.

The narrowness of the streets limited the construction of public and private gardens. Narrow streets were covered by arches and balconies of the upper floors. These balconies belonged to buildings that, contrary to the regulations of the time, were projected beyond the floor's front line (Fig. 4.2)

Notre Dame, the widest street in Paris at the time that was allocated to religious movements was only 7 m wide (Fig. 4.3). In other cities, the width of the streets was 1 to 2 m (Geisst 1997).

Fig. 4.2 Passages in a Middle Ages city; Siena, Italy. Authors' Personal Archive, 2019

Fig. 4.3 Paintings of Notre Dame in Paris in the Middle Ages. *Source* Griffintown (2010)

In 1185, the first clearing of medieval streets began. This also spread to Italy and France. In the fourteenth century, all streets in Troyes, Amiens, Douai, and Dijon were paved with stone. In medieval European cities, since street naming was not common at that time, flags were used as an effective element in marking passages.

At times, when the flags invaded the streets and prevented light from entering the houses, the governments tried to pass laws restricting their installation. However, even regulations for flags could not stop them from spreading. The issue of lighting was an acute problem for the cities of the period. At the beginning of the fourteenth century, the capital had lighting only on three streets. Occasionally, small churches and shrines built on sidewalks or street corners lit up the dark nights of these cities. It was only during the reign of Louis XIV that major advances in lighting were made. In medieval cities, trees were rarely seen in the streets and squares. The number of statues and fountains was small, and they were mostly luxurious. During this period, streets and squares were temporarily decorated for religious ceremonies (Habibi 2014).

4.2.3 Renaissance

In the seventeenth century, public spaces appeared in some cities that consisted mainly of garden alleys and public gardens. Marie de' Medici opened the 1500-m Cours la Reine, which was lined with elm trees, to the public in 1616 in Paris. This was the first avenue lined with trees.

Louis XIV and Colbert added the Cours de Vincennes, Les Grands Boulevards, and Champs-Élysées to the aforementioned street at a later date. People began to frequent these streets for leisurely strolls. The issue of street lighting was not neglected, and great progress was made. In 1667, for instance, La Reynie, the Lieutenant General of Police, issued an edict requiring the installation of 6500 candlelights in Paris. From 1700 onward, these lights were suspended in front of houses using ropes and pulleys. Originally, they were set on the windowsills by the homeowners. During this time period, the most important cities in France were equipped with exceptional lighting networks. In 1745, mirrored lantern lamps were developed and candles were replaced with oil. From 1769, all Parisian streets were equipped with these lights, which were positioned in the middle of the street at a height of 5 m and at intervals of 60 m.

During this period, plants were also used to decorate buildings. They were planted behind houses and, while creating a beautiful environment, gave the building facades a unique appearance. During this time, public parks and natural landscapes gained popularity in urban areas. In this way, Paris was an example for other cities. The city's walls were demolished and rebuilt with expansive boulevards (Geisst 1997).

4.2.4 Industrial Revolution

Beginning in the nineteenth century, urban spaces began to undergo significant changes. This period had a significant impact on the growth and success of these areas. This period saw the creation of alleyways in their modern shape. In the industrial city, churches and altars that were once scattered in the streets were removed. Additionally, the lighting system was greatly upgraded, and house and street numbers were assigned. During the nineteenth century, regulation and management institutions dominated the city. Haussmann enhanced the number of canopies and kiosks along the wooded streets over the same period (Shakibamanesh 2008).

4.2.5 *Twentieth Century: The Preference of Functionalism Over Aesthetics*

In the early decades of the twentieth century, "modernism" altered the appearance of streets, along with their spaces and constituent materials. Due to a growth in the number of automobiles and the resulting congestion on the streets, pedestrians no longer believed sidewalks to be a safe haven. Cities employed protective devices such as metal railings, monolithic metal fences, concrete bars, and short concrete walls to prevent vehicles from entering pedestrian areas and endangering pedestrians.

In 1965, the first pedestrian crossing was constructed in Rouen, France, concurrently with the first traffic plan. The approximately 300-m-long crossing was subject to various complaints. However, only two years after its implementation, the initiative has deemed a success and reproduced across France. The 1970s were the most advantageous decade for the development of high-quality urban areas in French cities. During this time, numerous transit plans emerged in various cities (Shakibamanesh 2008).

In 1976, the use of art in urban spaces, especially in city streets, increased, and attempts were made to establish an interaction between artistic expression and the people who were not accustomed to receiving it (Geisst 1997). In this chapter, we will focus more on the issue of streets in the new era, with an emphasis on the main streets.

4.3 Expert Perspectives on the Concept of Street

The quality of a city's public territories, i.e., its main streets, boulevards, parks, squares, plazas, and fountains, determines its attractive mental image (Paumier 2004). The streets are the flowing rivers of urban life and are where we congregate and travel toward the city's center (Whyte 1988).

The street is the most basic element of urban form. It is a physical entity that combines not only the buildings that define it but also the surrounding community. As an element that connects the settings of cities and towns, streets provide channels for movement and places for the development of the entire community. A variety of societal hues are presented across the streets. In actuality, it is the smallest component and form of the city at the city or town scale, although it contains the most intricate human life system. A street may be so intricate that it requires a lifetime to fully comprehend. They are particularly significant due to their capacity to link and accommodate diverse social groups. They function as an intermediary stage that merges multiple layers of public places with regard to human privacy or publicness, so serving as a communication tool and establishing a common ground for the private and public parts of human life. The classification of streets varies based on location, use, personality, quality, age, etc. There are streets throughout the world that are more prosperous than others, or even streets that evoke the history of the city in which they are located.

As Rob Krier puts it, the street is the result of the decentralized formation of houses and a land distribution structure that allows access to individual plots. It has more robust functional characteristics than a square. However, due to its size, a square is a more desirable place to spend time than a street, where people are forced to interact with the noise of passing vehicles. The only persons who can see the architecture behind the street scene are those who pass by (Krier and Rowe 1979).

In response to the topic of whether a street is an interface or a boundary, Lang (2006) asserts that streets serve as both an interface and a boundary, linking and separating distinct parts of the city. He argues that the perception of a city's quality is more dependent on the quality of its streets. The character of city centers is also dependent on the length and width of the blocks (width of sidewalks and car pathways, building heights, setbacks, number of entrances to buildings, presence or absence of shop windows, etc.). The quality of a street is also affected by the speed of passing motor vehicles, the manner in which cars are parked, the uses of the first floor on both sides of the street, and the street's flooring and furnishings. In stark contrast to modernists who turned their backs on urban streets, urban design in recent years has emphasized the characteristics of urban streets. In the same manner, as the Haussmann-era boulevards of Paris 150 years ago, many modern urban design projects follow guidelines for the placement of buildings along streets, as well as their use and (if necessary) setbacks (Lang 2006).

Jacobs (1997) discusses the importance of security in urban streets and believes that if streets are crowded and unpleasant, the city will be perceived as such and if the city streets are free from violence and fear, the city shall be safe and secure accordingly. In a successful urban area, even though the streets are crowded with strangers, residents feel safe and confident. People need not be frightened by many incidences of violence on city and urban region streets. In reality, fear of the streets makes individuals less willing to use them, which raises the perception of their danger. In reality, a busy urban street is more likely to be secure than an abandoned one (Jacobs 1997).

According to Jacobs (1997), a city street that can accommodate strangers and keep people safe should have three primary characteristics:

(A) There must be a firm separation between public and private space. As is commonly observed in suburban contexts, public and private spaces should not be mingled.
(B) There should be eyes on the street, the eyes of the street's natural owners. Instead of turning their backs or blank sides on the street and leaving it blind, street buildings that are suited to manage strangers and protect the safety of residents and strangers must face it.
(C) Sidewalks must be regularly utilized and incorporate activities that attract people. Nobody enjoys sitting next to a window and observing an empty street. Occasionally, large numbers of individuals amuse themselves by observing street activities.

4.4 Review of the Concept of Main Streets

Among various routes in a city, the design of its main streets requires high sensitivity and elegance because they are the main places where social interactions and urban activities are greatest both quantitatively and qualitatively, filling the minds of people with collective memories and shared mindsets about the nature of civic life. Main streets are full of many varied events, and people attend them for different reasons and at different times. Visiting significant activity centers, shopping, seeing shops and retailers, roaming and walking, and seeing others and being seen by them have transformed major thoroughfares into congested spaces with a variety of collective and individual activities for people. These streets are comparable to cities in that they are capable of presenting an unpredictable image and offering viewers a unique sense of excitement and activity at any time. Consequently, individuals' mental images of main streets consist of a dense concentration of events and a variety of urban life opportunities (TGM 1999).

The main street is the center of the community and is rich in history and character. Residents refer to it as the city center since it serves as the hub of urban life. As the oldest section of the city, the main street has historical significance and is a pleasant area to walk. Typically, it is an economically significant commercial center for the surrounding area. Importantly, the city's main street attracts people and a variety of commercial activities.

The main street is not restricted to the buildings that comprise it; it is the site where human endeavors, the evolution of architectural styles, the occurrence of many business operations and social changes, and the evolution of street design over time are inscribed. As a designer, Jacobs asserts that streets are constructed for symbolic, ritual, political, and social functions in addition to movement and access to other places (Jacobs 1996). The interconnected network of main streets, which is the result of years of effort by inhabitants to support urban life, is filled with their memories, preoccupations, needs, and preferences. It reflects the lifestyle, mindset, economic status, type of social relations, and other subjective and objective commonalities of the city's occupants.

The vitality and civic life of main streets are not the consequence of unidirectional design and construction. Therefore, a functionalist or formalist approach to design and construction cannot be relied upon to expedite their development (Pakzad 2013). In fact, the most notable characteristic of main streets is that they are the result of gradual growth and a natural process of change in the city. Rather than reflecting the physical structure of the city, these streets represent the social, economic, and cultural structures of society. Thus, neither can the status of the existing main streets be reduced to that of a "passing artery" under the pretext of resolving the issue of automobile traffic, nor can they be altered and decorated artistically and intuitively, erasing their collective memory with self-centered decisions. The main street reflects the civic life of a city over time. In developing new cities or areas envisaged for city expansion, it is therefore physically impossible to design and construct the main street all at once. Through using scientific and empirical methods to predict the socio-economic context of future residents (interests and needs of future citizens), one can at best hope to gain some inspiration for planning and designing routes that will become the main street after a relatively long period of time, based on the integration of the commonalities and collective needs of the citizens. Over time, like a string of rock candy, it absorbs the potentials of its surrounding surroundings and is eventually saturated with all available potentials (Pakzad 2013). This does not imply that other urban axes will have all expected aspects of public life immediately after design and construction and will not be affected by the passage of time, but it does imply that main streets are distinct from other types of urban axes due to the number and variety of events that occur on them, as well as other similar distinguishing features that are more rigorous and fundamental within them. Because the quantity and types of events, the diversity of social groups, and the role and intended functions of places are more limited and particular in other city streets, designers, and urban planners have more possibilities for the design, construction, and operation of these streets.

A community's sense of place is manifested in the main streets of city centers. Many communities have come to the conclusion that they have lost their sense of place. Whatever the reason (e.g., economic recession, changes in land use, especially strip development and large-scale shopping malls, and sometimes adverse changes in access systems, etc.), these communities have started efforts to regain that sense of place (TGM 1999).

Main streets are made up of several blocks in length and width that are usually dense and include mixed uses. The buildings are usually adjacent to each other and to the street. Main streets have narrow blocks and are connected to the local network, including street parking lots. They have sidewalks wide enough for two people to walk side by side. These streets are pedestrian-oriented in terms of traditional texture and design principles. Today, high-speed pathways in contemporary cities are inevitable, but at the same time, care must be taken not to turn the main street into a crossroad, destroying urban life and the intertwined network of urban life. Organizing and/or restoring main streets requires great care and time as well as extensive facilities, and overlooking them can deprive a generation of having public arenas for collective life. The mistakes made by our experts and city managers have turned the experience of main streets into an amalgamation of car traffic, pedestrian movement, chaos, noise, and pollution that are perceived as synonymous with disorder and confusion by users. These streets, which form the essential structure of cities, have been suffering from increasing problems so that people are only willing to appear there to satisfy their personal needs.

4.5 Past and Present of Main Streets: From Prosperity to Decline

Main streets often emerged as city centers along the most important (and sometimes the only) pathway. They grew alongside the development of the city, but the original main street sometimes might progress only a few blocks among several kilometers of new development and construction.

Initially, the main street facilitated access by shortening roads (the principle of proximity) and helped locate required uses such as municipalities, post offices, grocery stores, banks, doctors' offices, shops, restaurants, and leisure centers. Even though the initial main streets were surrounded by huge farmland and orchard plots, their primary purpose was to provide access for people and automobiles to the city's services and commercial activities. One-stop shopping remains a significant benefit for many main streets (Moe and Wilkie 1997). As building patterns moved to accommodate automobiles, the street's emphasis shifted from "accessibility"[1] to "movability".[2] This shift has been emphasized over the last 40 years, and the prioritization of mobility within city limits over the accessibility of uses has resulted in a decline in the core values of Main Street. There is an immediate need to find a balance between "accessibility" and "movability" on main streets given the current scenario (TGM 1999).

In the past decade or so, the conditions in our cities, particularly in our metropolises, have elevated access networks and routes that have a direct impact on a variety of city concerns. This phenomenon describes the nature of "human presence" on main streets to a considerable extent. On the qualitative side, the factor of "presence" appears to be complex and multi-layered, despite the issue's apparent simplicity and quantifiability through field research (Tabande 2004). Due to the cultural and civic roles that they might play in an urban setting, the presence of people on the main streets is of great importance. Such roles have been badly impacted by numerous issues, the most significant of which is the collision between automobile and pedestrian movement. Today's main streets are never comparable to the concept of boulevards in the true sense of the word (as seen in the Chaharbagh in Isfahan or the Champs Elysees in France), not only in terms of comfort and living but also in terms of the most obvious and primary environmental conditions. The streets of those eras were a symbol of culture and power, while the streets of our cities today are just a traffic convolution.

As mentioned above, the term "main street" refers to the city center or commercial areas within towns or cities. They are often surrounded by residential areas and serve

[1] • Accessibility refers to the ability to access different goods, activities, and destinations. The time and cost of travel determine the quality of access. Accessibility is influenced by individual factors (wealth, physical ability, age) and social factors (land use patterns, quality of public transportation, walking conditions, capacity of roads, telecommunications, etc.).

[2] • Movability refers to the ability to move and the availability of pathways to access a destination. Mobility may be provided by vehicles (buses, cars, bicycles) or on foot. Increased mobility does not necessarily indicate better access, as travel costs and time may also increase.

as a place for community members and people from neighborhood units. In the past, main streets were places where most of the city's activities took place and served as a nucleus for transportation, shopping, meetings, religious and civic activities, government presence, and business (Moe and Wilkie 1997).

In "Changing Places: Rebuilding Community in the Age of Sprawl," Moe and Wilkie (1997) describe main streets as follows:

> After the Second World War, the main street became a unique business center as well as a cultural core for urban communities. The main street hosted the heritage of previous generations, embodied in monuments or public investments in the form of sidewalks, streetlights, etc. It was a place for the manifestation of the local government and its institutions, as well as for holding formal and informal public gatherings, holiday ceremonies, and annual celebrations. The main street was a place to shop, do business, get entertained, and exchange local news in theaters and restaurants, on park benches, at lunch tables, and in cafes, barber shops, and beauty salons. There were many people on the main street, mainly because a large number of them lived in the neighborhood, in a square in a residential street, or in a cheap apartment above a shop or office within a short distance from the essential places in the city. The main street was crowded at all hours of the day. But in just a few decades, support for these urban spaces faded away. (Moe and Wilkie 1997)

Longstreth (2000) writes:

> Main Street was one of the first public places in the United States. … in fact, main street in the United States was equivalent to main plaza in Italy. (Longstreth 2000)

However, today, most main streets lack such qualities, as they are no longer able to create the vibrant and dynamic atmosphere as before. Main streets, including those in cities and towns, have lost their integrity as a result of various environmental, political, economic, and social factors and no longer have the experience they used to create for locals or visitors. In fact, main streets, especially in developing countries, are declining and lack the potential to strengthen social interactions, create an identity for the region, or play a role as a social, civic, religious, and commercial center. Fewer people now live on main streets than in the past because the environment is no longer a valuable place for the surrounding community. Hence, the main streets of the city are currently in disarray. The related city contexts, which have been destroyed for the construction of parking lots or new streets, are now fragmented and have lost their integrity. Many buildings are empty, and some are falling apart. Private or public sector investments are also very small, and sometimes the funds needed for the revitalization of such spaces are not provided at all. Most residents feel irresponsible about their place of residence, and as a result, they are forgotten just like their environment.

The issues mentioned above as well as main street properties like memorability and vitality and their role in the whole set of environmental qualities related to the city show the importance of main streets and the need to pay attention to different dimensions of their design, including the social dimension (the centrality of collective interactions), the aesthetic and visual dimensions (evaluating the qualities and aesthetic experiences of the streets and promoting these experiences), and the functional dimension (the centrality of urban life and the importance of their vitality).

4.6 Revitalization and Reconstruction of Main Streets of City Centers

4.6.1 The Necessity to Restore the Main Streets

Today, many main streets are unable to support and provide social activities. Due to the fact that a community requires a place for people to gather, the absence of such places will make it extremely difficult for a true community to form. Creating a place where community members feel a sense of belonging is crucial to re-establishing the concept of community and a sense of civic pride. The revitalization of main streets creates true urban spaces by providing a venue for incidental meetings and face-to-face relationships between residents.

The main streets of cities require revitalization and restoration. They consist of irreplaceable physical structures that convey the history and character of a region. Additionally, old buildings and areas, which should be similarly protected, can unite existing communities through a sense of belonging to the past and create a shared identity. Invigorating these places shows the high value we place on the past and demonstrates our understanding of history. In addition, it strengthens the local economy and enhances the dynamics of social interactions, resulting in the desirability of living in such communities (Longstreth 2000).

The acceptance of neo-urbanism for its efforts to revitalize main streets demonstrates that community members are still interested in a positive and powerful image of main streets in their vicinity so that these areas can continue to serve as symbols of city centers. Main streets are the physical manifestation of a community's sense of self and are a significant source of identity and social pride. Consequently, different main streets within a large city must each have a local identity that is consistent with the larger and more dominant identity of their contexts (Katz and Scully 1993).

The revitalization of the main street can serve as an environmental and economic stimulant in the region, as well as improve the visual and social status of neighboring areas by increasing investment and boosting economic and social aspects. This process of revitalization can begin in the commercial district of the city center and spread to neighboring residential areas. In addition, the revitalization of main streets and their surrounding areas can significantly reduce the number of vacant and unused spaces in the urban core. In most cases, a small town is created outside of the metropolitan area as a result of new urbanism-based developments intended to revitalize small towns, resulting in the destruction of farms and agricultural lands and a significant waste of resources. However, revitalizing existing main streets through infill development and enhancement of street landscape will conserve land, resources, and other assets.

Main streets in city centers should be activated by means of refined urban design, controlled architectural intervention, and government financial support. These interventions should preserve the existing context, which contributes to the formation of the character of the main street, while also enhancing the environment to meet the current needs and requirements of the associated community.

The factors that led to the decline of main streets must be addressed, as well as current issues in technology and transportation. The revitalization of main streets in city centers cannot be based solely on an architectural strategy. Therefore, existing theories such as new urbanism, urban renewal, critical regionalism, and preservation should be combined to create an all-encompassing theory of main street revitalization. Despite the fact that each of the aforementioned theories is independent in and of itself, specific components of each can be extracted and redirected to achieve a more comprehensive approach to main street revitalization.

In this regard, new urbanism can be utilized as a strategy for arranging spaces and structures in traditional areas. Regionalism can be advantageous for the architecture of urban spaces. The concept of urban renewal should be modified to create sites for new construction, and the theory of conservation can be used to revitalize existing structures. In the sections that follow, these methods are discussed in detail.

4.6.2 Theories of Main Street Revitalization

(A) **New Urbanism**

New urbanism is an approach that emphasizes the design of successful and conventional neighborhood units. It is premised on the notion that existing units should be renovated first, but this has proven to be economically, politically, and sometimes even physically impossible (Katz and Scully 1993). According to Fulton (1996), new urbanism follows two major goals:

- to foster a stronger sense of community in neighborhoods by allowing for a greater diversity of land uses and social interaction, as well as by rethinking the "public realm," especially public spaces and the typical streetscape."
- to reorient the typical community toward a pedestrian- and transit-friendly environment, while also minimizing negative traffic and environmental effects" (Fulton 1996: 3).

To realize and attain these objectives, new urbanists have drafted the New Urbanism Charter, which explains the requirements of development based on historical and traditional neighborhood units. The charter provides physical design-based solutions to problems such as transportation, optimal land use, zoning regulations, affordable housing, and the dearth of a dynamic urban community.

New urbanists apply the principles that make a city successful in its activities. These principles are derived from The Death and Life of Great American Cities, a 1960s book by Jane Jacobs that describes various aspects of a successful city, including the following criteria (Jacobs 1997):

- The district and the majority of its internal parts should have multiple uses, preferably more than two types, to ensure the presence of individuals with diverse intentions and schedules who can utilize numerous shared—facilities.

- Most blocks should have a narrow street-level width. In other words, there should be a high number of streets and opportunities to turn corners.
- The district should consist of a variety of building types, ages, and conditions, including a significant proportion of older structures, in order to generate a variety of economic returns. The composition should have a relatively dense concentration.
- There should be a sufficient gathering of people with different goals on the street. This also includes a dense concentration of people who are there because of residence in the district.

The advantages of new urbanism include mixed-use zoning and planning, 24-h activity zones, and pedestrian-oriented planning that creates a maximum 5-min walking distance between locations within an area. These are among the most significant elements extracted from the works of Jane Jacobs and her predecessors to formulate the principles and frameworks for new urbanism-based development and construction.

Duany et al. (2001) list six design criteria for new urbanists based on the distinction between traditional neighborhood units and suburbs.

- **Neighborhood Center**: Commercial, civil, and social uses can be clustered in a central area, such as the main street.
- **The Short Walk**: A five-minute maximum walking distance between houses and neighborhood centers should be considered.
- **Dense Street Network (or small-scale blocks)**: This creates multiple routes for automobiles and pedestrians to travel throughout the district.
- **Narrow Streets**: They create an environment commensurate with the movement of pedestrians and reduce the speed of automobiles.
- **Mixed Use Buildings/Spaces**: They help to create a district with 24-h and diverse functions.
- **Unique Sites for Important Buildings**: Including civil structures (such as municipality).

It is difficult to apply the principles of new urbanism to every building in an urban context because no architectural style is justified by the principles. However, the majority of new urbanist buildings have a traditional design, and urban planners can provide architects with design guidelines based on the available styles.

New urbanists believe that the visual integrity of districts is the most important factor in the design of buildings. Therefore, they follow the traditional design style and principles of the districts. New urbanism seeks to find the familiar mental image of a district. Proponents of this strategy believe that returning to the past and its values will satisfy the needs of individuals and induce a feeling of nostalgia. Under conceptual and classified conditions, new urbanism also seeks to control the environment. According to new urbanism, a "place" is created by the familiarity of the environment, mixed uses, and well-designed regional planning. According to this school, individual buildings are not built in the same way or following the same method as traditional buildings in the past. Even though they are constructed with current measures and facilities, they are decorated and adorned according to past designs. Some critics of new urbanism see this as the creation of a false layer. However, new urbanists believe that since buildings create a special visual whole, forging old decorations into them is acceptable (Katz and Scully 1993).

According to new urbanists, planning regions in accordance with their principles creates places and economic incentives that permeate the entire district. The Charter of New Urbanism and its other tenets can be used to design major streets. Specifically, these principles are founded on existing regions and circumstances. Recent viewpoints indicate that main streets should not necessitate an integrated, similar-principles-based environment. In many instances, it is not necessary to define a specific context for a street, because the main streets of a city can be a collection of architectural styles from the past and present. However, this strategy can be utilized effectively, particularly in traditional areas with their own distinct character, where the majority of the spatial layout and design have already been completed. By providing effective guidelines, the Charter of New Urbanism can aid in the development of community-oriented areas.

(B) **Regionalism**

While new urbanists describe the factors that influence the success of neighborhood units, regionalism examines and defines the conditions under which a building or area becomes a place. It is implemented in areas with simple and limited contexts, but the vast majority of its projects are intended to complement an urban fabric and fill empty spaces (infill development). In reality, globalization leads to the gradual unification of urban environments in various countries, despite the fact that traditional construction methods, history, and people are clearly different. Regionalism opposes imitation and homogenization and affords numerous opportunities for various districts to have unique designs. This school explains the need for a place while distinguishing the issue of similarity from that of specific environmental characteristics.

In fact, regionalism is a movement that seeks to develop an architectural style that reflects the cultural, economic, and political characteristics of a particular region. However, it is not intended to produce local architecture. Instead, it attempts to achieve a balance between local, national, and international building features (Cowan 2005). In addition, regionalism and a nostalgic approach should not be confused. In accordance with this strategy, the form and character of the space should be determined by traditional and indigenous components and principles. Thus, regionalism is an effort to reinterpret the actions and principles of traditional architecture in a particular region so that they are compatible with the contemporary environment. Examining the traditions of indigenous and local architecture aids in defining the character of a region. This school strives to provide a unique response to the culture, geography, and climate of each region. Nonetheless, regionalism can deteriorate to the point where it becomes a superficial historical insistence that combines images of regions with national and cultural stereotypes. At best, regionalism reconstructs the symbolic principles and infrastructures of the past and adapts them to the prevailing variable social order (Tzonis and Lefaivre 1996).

Traditions are susceptible to change and transformation under regionalism. Consequently, this strategy rejects the acceptance of a tradition comprising a fixed collection of tools and images. Regionalism refers to an architectural style that reflects regional traditions, forms, construction methods, and history. However, the region does not necessarily correspond to a particular geographic area. It can refer to diverse races and groups, common climatic and geographical characteristics, political boundaries of ethnicity, and other classifications. According to this viewpoint, a city's main street can also be classified as a region due to its similar traditions and culture. By examining the history of main street and the development of traditional building styles, we can formulate and define particular principles, and by transforming and modifying some of these traditional principles, we can create a system based on regional principles. Also, in many cases, given the specificity of regionalism, it must be used alongside another urban approach to revitalizing an area.

(C) **Urban Renewal**

The concept of urban renewal was inspired by modernist ideas. The movement is sometimes associated with a negative connotation, as it is blamed for the destruction of numerous delicate and valuable urban contexts, particularly during the middle of the twentieth century. It was an attempt to restore cities' middle classes by removing elements of the urban environment that were unattractive to the people. Even though the number of urban renewal projects is significantly lower than it was in the 1960s, there are still cases of such projects because they are less expensive and are still viewed by some politicians as a symbol and indicator of growth.

To understand the history of urban renewal, one must first realize modernism. Modernism originated in Europe prior to World War II as a reaction to the rapid expansion of industry and industrial production. Its architecture could be distinguished from those of other movements by avoiding historical decoration and symbols. The status of modernist structures was equivalent to that of historical buildings, but they were constructed from steel, concrete, and glass. Modernism sought to create a new type of architecture for the new generation of industrial men. According to Kunstler (1993) in "The Geography of Nowhere," modernist architecture is clean, functional, safe, healthy, and bright, with separate habitats where the new industrial man can build a new and superior culture away from his past mistakes. In post-World War II Europe, modernism was used for reconstruction. This movement appeared sufficient and offered quick solutions to problems in war-ravaged areas. In the United States, particularly during the 1960s, when cities were in dire need of reconstruction, modernism and modern planning were utilized to repair and reform cities. After World War II, modernist strategies were also employed in a number of European cities.

In the aftermath of World War II, the United States began to interpret modernist ideas and apply them as principles of architecture and urban planning in the process of urban renewal. Instead of restoring worn-out urban areas, the majority of them were demolished and vacated in order to create an environment that people consider to be better than it was previously. For the new social and industrial movement, it was believed that old neighborhood units were not a suitable base. As a result, rather than undergoing a successful restoration process, which is currently the most effective step in revitalization, new buildings were constructed on the old sites because it was more expedient. It seemed reasonable at the time to demolish existing neighborhood units in order to facilitate growth and development. Politicians viewed contemporary structures as symbols of expansion. The fact that modernism was more affordable was also welcomed by developers. The movement attempted to reintroduce the middle class to the city, an endeavor that was successful. Modernism resulted in the construction of roads between cities, suburbs, and other regions, as well as the construction of large office towers for suburban residents in cities. The delicate balance between revitalizing neighborhood units with identity and human scale, on the one hand, and renewal projects, on the other hand, has been lost. Numerous areas desperately requiring restoration were obliterated (Moe and Wilkie 1997).

Due to the negative connotation associated with the concept of the past, the term "urban renewal" is no longer used to describe urban renewal projects. However, there are still instances in which the entirety of an area must be demolished in order to construct a larger and more comprehensive structure. Most alternative buildings are not considered modern elements, but rather buildings that are more context-oriented and sensitive to their surroundings.

(D) **Preservation**

In his definition of preservation in the book Changing Places: Rebuilding Community in the Age of Sprawl, Richard Moe states that protection, interpretation, and improved understanding of works or features that are historically or architecturally significant are always at the heart of preservation. The movement is founded on an appreciation of the significance of history, but it is not preoccupied with the past. It is the art of maintaining specific places and the quality of life they create and sustain. Thus, something other than bricks, columns, decorations, and stones is highlighted (Moe and Wilkie 1997).

The purpose of preservation is to prevent the destruction of existing environments and buildings. Priority is given to maintaining existing buildings in an environment with a focus on revitalizing the area's economic, social, and environmental conditions. Because they serve as symbols and manifestations of power and identity, preservationists view the preservation of historic buildings and areas as crucial. In fact, both old and new buildings reflect the passage of time. In contrast to living in a place where everything belongs to the present, preserving old buildings allows people to have a strong sense of identity, measure their own and their predecessors' time of residence, and evaluate the lives of various buildings in comparison. Additionally, preserving contexts from various time periods enables individuals to read and comprehend the history of particular areas. On the other hand, the preservation of buildings allows for the development of unique uses for them. However, if the preserved structure cannot be used for any acceptable purpose, a new structure must be constructed in its place. In addition, preservation is an effort to prevent the neglect or destruction of a place or region during the development process. For instance, areas with a rich history that lack thriving economic conditions or are incompatible with the architectural preferences of the present are at risk of being destroyed.

The preservation movement preserves these areas and uses them as a foundation for environmental and economic development.

The movement's adherents believe that the preservation of historical elements in an environment that promotes a positive and socially active lifestyle is both desirable and essential. We must protect the existing environment for the four reasons listed below (Moe and Wilkie 1997):

- **The economic dimension:** Given the significant investments made in previous communities, whether in buildings or public infrastructure such as streets, water supply facilities, sewage systems, and so on, wasting them is financially irresponsible and irrational.
- **Understanding history and the past:** Through preservation, people will be able to better understand themselves and the society in which they live.
- **A better response to the general needs of urban society**: In fact, traditional areas better meet the needs of public life than suburban developments.
- **Preventing inner-city areas from getting forgotten and destroyed:** Without preserving the inner-city areas, people will continue to migrate outside cities, leading to the destruction of the inner parts of the city.

In general, the preservation approach creates a common identity for people in an urban area. This attitude also contributes to true place formation. However, to be useful for main street, which is a collection of abandoned spaces with disproportionate architecture inconsistent with the background, the approach must be used in conjunction with an urban theory and a new construction theory.

References

Cowan R (2005) The dictionary of urbanism. Streetwise Press

Duany A, Plater-Zyberk E, Speck J (2001) Suburban nation: the rise of sprawl and the decline of the American dream. North Point Press

Fulton W (1996) The new urbanism: hope or hype for American Communities? Lincoln Institute of Land Policy

Geisst CR (1997) Wall Street: a history. Oxford University Press

Griffintown (2010) Historical documents: Notre Dame Street in 1850. Retrieved from: http://griffintownyesterdayandtoday.files.wordpress.com/2010/12/d121.jpg

Habibi SM (2014) Azshar tashahr; a historical analysis of the concept of the city and its physical appearance: thinking and impression. Tehran University Press

Jacobs AB (1996) Great streets. The MIT Press

Jacobs J (1997) The death and life of great American cities. Cape

Katz P, Scully V (1993) The new urbanism: toward an architecture of community. McGraw-Hill Professional

Krier R, Rowe C (1979) Urban space. Academy editions

Kunstler JH (1993) The geography of nowhere: the rise and decline of America's man-made landscape. Simon & Schuster

Lang J (2006) Urban design: a typology of procedures and products. Routledge

Longstreth R (2000) The buildings of main street: a guide to American commercial architecture. Altamira Press, New York

Moe R, Wilkie C (1997) Changing places: rebuilding community in the age of sprawl. Henry Holt & Co., New York

Pakzad J (2013) Urban space design guidelines in Iran. Ministry of Housing and Urban Development Press

Paumier C (2004) Creating a vibrant city center: urban design and regeneration principles. Urban Land Institute

Shakibamanesh A (2008) Promoting city centers quality by using special urban design guidelines (master's thesis). Iran University of Science and Technology, Tehran, Iran

Tabande R (2004) Street symphony. J Urbanism Essays (13)

Transportation and Growth Management (TGM) (1999) Main street handbook. Oregon

Tzonis A, Lefaivre L (1996) Why critical regionalism today? In theorizing a new agenda for architecture, an anthology of architectural theory. Princeton Architectural Press, pp 209–217

Whyte WH (1988) City: rediscovering the center. Masscases.com. Retrieved from http://masscases.com/cases/sjc/427/427mass509.html

Chapter 5
Principles of Designing Successful City Centers

5.1 Introduction

It is obvious that successful city centers in the world have not achieved such success by chance. They have also not followed a merely stereotypical plan. The purpose of this chapter is to formulate, summarize, and present solutions that can lead to the creation of a quality, vibrant, and dynamic city center and to provide a more comprehensive functional package for city center design. In this review, fifteen principles for transforming the city center into a successful public space are presented. These principles and guidelines are presented in such a way that they can be implemented in cities of varying sizes. These principles should guide the decisions that shape the form and personality of the city center as a quality place for people and can be used to evaluate the available city center facilities. These principles consist of:

- Development based on an organizing pattern.
- Formation of a distinguished identity.
- Creating diversity and interest.
- Ensuring the balance of activities.
- Creating a city center based on local facilities and resources.
- Eliminating incompatible activities and uses.
- Ensuring visual and functional continuity.
- Maintaining relative enclosure with human scale.
- Maintaining city centers' concentration and compactness.
- Increasing accessibility and pedestrian and vehicle access to the city center.
- Security and safety of city centers.
- Comfort and well-being.
- Design based on different tendencies and tastes.
- Emphasis on high quality.
- Balance between the static and dynamic aspects of pedestrian movement.

In this chapter, these design principles are discussed in full detail.

A. Shakibamanesh and M. Ghorbanian, *Designing Responsive City Centers*, Design Science and Innovation, https://doi.org/10.1007/978-981-97-8911-5_5

5.2 Principle 1: Development Based on an Organizing Pattern

A simple and clear pattern of development in the city center can provide residents and visitors with a framework for comprehending the organization of an urban area and navigating within it. Reduced distances between buildings and increased heights, for instance, should provide a visual guide for the physical location of the administrative core of the city center. The primary public spaces should designate pedestrian zones in the city center. There should be a concentration of retail establishments in one area, adjacent to the administrative center, and other areas should be located based on their performance. In this manner, each area will develop its own character and interact physically and functionally with neighboring areas to form an integrated city center (McBee et al. 1992).

This organizing structure is crucial for the formation of an identity and a special and distinct sense of place in the heart of the city. A legible development structure assists individuals in locating themselves and identifying the functional-activity core of the city. It creates a city that helps people find their way, facilitates access to the city center, and fosters a sense of physical and psychological security, comfort, and desirability. In The Image of the City, Lynch (1964) discusses the "reading" of urban environments and the need to arrange its various components and parts in a coherent and logically connected pattern. The book provides urban designers with a strong empirical base for making practical decisions. The usefulness of Lynch's theories regarding pathways, edges, areas (neighborhoods), nodes, and landmarks has been demonstrated to date. Various arrangements and solutions can be used to create a cohesive organizing pattern for the city center, as discussed below.

- The pattern of streets and blocks in the city center can significantly contribute to create an organizing structure in the area. The organizing power of a structure can be determined by factors such as the size of blocks, the order and number of spatial intersections, the pattern governing the network. For instance, blocks of relatively short length can limit the width of buildings, resulting in a greater visual-functional desirability for pedestrians by creating connections with other blocks and sub-areas and enhancing physical and visual permeability. In contrast, incorporating a hierarchy into the design of thoroughfares can enhance the visual structure of the city center. By reducing the likelihood of collisions and intersections with local traffic, it can also increase the efficiency of the traffic system, decrease traffic volume, and slow down vehicles in city centers.
- Strong visual cues can strengthen and reveal the functional organization of a city, thereby enhancing the inhabitants' experiences in this regard (Fig. 5.1). Reduce the distance between buildings and increase their heights in order to direct movement toward the city's core. Similarly, the main cues (valuable elements, significant public and open spaces, etc.) should mark the intersections of roads and streets in the heart of the city center. By creating highly visible landmarks, open spaces can improve the legibility of a city center, for instance, if they are strategically placed in the general street and block layout. These open spaces provide a focal

Fig. 5.1 Some buildings in the city center serve as strong and directional visual cues for people, particularly pedestrians. Bologna, Italy. Authors' Personal Archive, 2019

point upon which urban development and diverse land uses can be centered. These spaces situate small developments within the context of the city center, creating a unified and attractive whole.

- The way constructions relate to the street can reinforce or weaken the clarity that reflects the integrated and interconnected structure of the city center. For example, construction with a proper and defined density and scale can indicate the entrance to the heart of the community. Conversely, if the construction along the main entrance to the city center is uniform in use, scale, and degree of importance (like linear commercial developments), people will become confused about where the city center starts and ends and how it relates to the surrounding areas.
- Within the city center, the buildings play an important role in reinforcing the organizational pattern by creating a continuous and solid architectural edge that defines the street space (Fig. 5.2). When empty lots, surface parking lots, and deep setbacks of buildings on the streets create long empty spaces on the street front, the three-dimensional framework of buildings and streets is weakened. On the contrary, regular, consistent setbacks of yards in front of buildings and infill developments among structures that preserve and strengthen the urban context can unify the structure of the city center (Paumier 2004).

In conclusion, these are the most effective elements and pillars of the city center's public environment for creating an organizing structure. They include main street patterns, the role of streets in the traffic hierarchy, street landscaping arrangements, the position and characteristics of open spaces, and the structure of the traditional market. Through public policies and investment decisions, these principles should be organized and managed to create strong urban design frameworks that can ensure structure, continuity, and coherence in the city (McBee et al. 1992).

Fig. 5.2 The continuous and solid architectural edge in the passageways leading to the city centers creates an integrated organizing pattern. Florence, Italy. Authors' Personal Archive, 2019

5.3 Principle 2: Formation of a Distinctive Identity

Over the past few decades, Americans have become accustomed to shopping malls and covered markets (see Chap. 3), which are artificial environments in which everything from the layout of parking lots to the design of storefront windows is strictly regulated. Given the predominance of traditional markets (Bazaars) in past in the urban cores of many Asian countries (such as Iran, Turkey, Iraq), this mental image was stronger and more distinct in Asia than the urban spaces and centers of other countries.

In fact, city centers must be characterized by distinctiveness and originality. Authentic places, not attractions, are what attract people. According to Pihlak (1988), we should create authentic and desirable places to attract locals to the environment. If they visit these places, tourists will follow. The inherent and existing characteristics of a city center, such as its location and historic buildings, can distinguish it (Pihlak 1988). A successful space, according to Abbaszadegan (2006), must communicate with users on three levels: individual, group, and community. At the individual level, communication should be designed to evoke a sense of continuity between different life stages. Sociologists sometimes define this type of communication in terms of "spatial identity" or "place identity." He believes that a person's positive relationship with space is crucial to the formation of a sense of "self" and "personal identity" (Abbaszadegan 2006). The preservation, revitalization, and reconstruction of old city neighborhoods where residents have individual and group relationships is an important design objective. Consequently, according to individual and group sociology and psychology, the destruction of a city's historical identity results in the destruction of a portion of the individual, group, and even national identities (Abbaszadegan and Rezazadeh 2002).

A space has different connotations for various groups. As a result, they relate to it on various levels. A street, an entrance, or an open space in the city center can have a symbolic significance and, as a result, a unique identity for a group of people or even the entire population of a city or country. In the urban design process, it is crucial to recognize these identity-related meanings and concepts as well as the type of relationship (sensory or experimental) people have with the identifying elements of the city center. For instance, "Nakhl" occupies a special place in the traditional city centers of Iranian cities and moves in space over the shoulders of Shiite believers to commemorate the high religious and human concepts. Even after the ritual has concluded, Nakhl remains in the space, imbuing it with a unique identity (Fig. 5.3).

In fact, city centers require a distinct identity to create an attractive and desirable environment that encourages human interaction. The coordination of cultural, recreational, and sporting activities with special events (such as festivals and mourning rituals) contributes to the formation of such an identity. Residence and improving life in the city center and setting up weekly markets and sale events are other factors considered effective to this end.

A vibrant and distinct mental image can imbue the city center with both personal and social meanings. A city center with an appealing identity can serve as an effective marketing tool for all city center uses. A pattern of city center construction and development that is supported by an orderly network of streets and building blocks with continuous edges and strong enclosing walls is a suggestion for creating such a clear mental image. Repeated use of micro visual elements, buildings, street views,

Fig. 5.3 Nakhl Gardani is a Shia religious ceremony performed on the Day of Ashura to commemorate the death of Husayn ibn Ali, the third Shia Imam. Nakhl is a wooden structure used as a symbolic representation of the Imam's coffin, and Nakhl Gardani is the act of transporting the Nakhl, which resembles an Imam's funeral procession. The ritual is performed in the historic centers of numerous Iranian cities. Yazd, Iran. Authors' Personal Archive, 2016

and open spaces in city centers can also create multiple layers of detail, resulting in a cohesion and consistency. This contextual cohesion may involve spatial relationships between architectural details and forms, materials, colors, paintings, street furniture, etc. While creating the desired image, the selection and layering of background elements must be carefully considered to prevent the inclusion of too much detail, as in designing Hollywood scenes. It is necessary to identify coherent and consistent contexts so that strategies can be developed to reinforce and enhance their coherence.

Where city centers are large enough to contain specialized subdistricts, multi-purpose centers develop. In such city centers, each subdistrict must have its own distinct identity while still being distinguishable as part of a larger urban core. Some design considerations (such as streetlights, pathway trees, pavement materials) can introduce the core and be applied to subdistricts as a whole. To differentiate the character of each subdistrict, it is suggested that minor modifications be made to these elements. It is also necessary to create safe, well-defined pedestrian crossings that connect each subdistrict to the city's core (Paumier 2004).

5.3.1 Historic Buildings

Human scale, high-quality materials, and intricate architectural details in buildings constructed in the nineteenth and early twentieth centuries in Europe and the United States, as well as during the Safavid and Qajar eras and even earlier in many Iranian city centers, are unique and potent sources of physical identity. Whether as significant urban indicators or as supporting structures related to the distinct character of the subdistricts, old buildings provide a sense of urban continuity or a connection to the past achievements of society. Future city center decision-makers and officials must confidently work to preserve the city's historic landmarks, buildings, and blocks in order to preserve this capital. Repairing old buildings, adapting new activities and uses, and restoring the facades of old stores can enhance the urban landscape and the city's image. Even actions as simple as repainting or replacing signs can improve the appearance of the city center and create a sense of vitality.

New construction must reinforce or complement the elements that give the existing architecture its character and attraction. This is particularly true on the streets, where visibility and density are greater. Strengthening local or regional architecture rather than using foreign architecture should be encouraged, as these local characteristics make it easier to form a vivid mental image of the city center (Fig. 5.4) (McBee et al. 1992).

Fig. 5.4 Protection of buildings with local and historical architecture in the central core of Dijon aids in conveying the city's identity and instilling a vivid mental image of the historical center in the minds of the visitors. Dijon, France. Authors' Personal Archive, 2022

Fig. 5.5 The unique natural environment, including diverse vegetation and the river near Bern's historical center, has endowed it with a distinct identity. Bern, Switzerland. Authors' Personal Archive, 2022

5.3.2 Geography

The natural location and conditions of the city center, such as its elevations, views, and other natural assets, can give the city and its central area a distinctive identity (Fig. 5.5).

5.3.3 Landmarks

Different types of landmarks include buildings, arcades, public spaces, fountains, and clocks. Landmarks are differentiated from other components of the city center image by their ability to stand out from their surroundings. This attribute guides users and provides them with a sense of civic pride and identity. Landmarks are the most memorable elements of the urban landscape and represent and symbolize the entire city center, particularly for out-of-towners. They can be used for branding and associated with a positive mental image, allowing them to become widely recognized.

Fig. 5.6 A variety of urban landmarks can be found in Bern's historic center. Locals and tourists find the fountains and statues that convey cultural concepts, folktales, and a sense of community to be extremely attractive. Bern, Switzerland. Authors' Personal Archive, 2022

Improving and promoting public spaces, particularly well-planned private developments, can result in the creation of new landmarks that serve to distinguish the city's core (Paumier 2004). Buildings that are distinguished from their surroundings by their height, density, architectural style, or abundance of components can also serve as useful landmarks. However, these elements are most effective when presented in contrast to a uniform and cohesive context (Fig. 5.6).

5.3.4 Street Landscape

Since streets are a prerequisite for movement in the city center, street landscapes can contribute to the city center's attractive identity. This can be accomplished through the use of well-designed benches, lights with special standard designs for streets, green spaces in passageways, and other street landscaping elements. By visually distinguishing the streets or altering the land use, distinct pavement materials can facilitate the integration of the central area. These elements should be designed and combined as a whole, not separately or without consideration of their impact on the public space's overall character. This can be the key to improving and maximizing the quality of the streetscape and creating a sense of identity in the city center that is consistent with urban design (Fig. 5.7).

5.3.5 Public Spaces

A public space situated at the intersection of city center thoroughfares can play a significant role in establishing a city's identity. If the space is both a strong visual

Fig. 5.7 The precise application and mix of street lights, trash cans, and standing signage, as well as the use of suitable flooring materials, have improved the quality of streetscape. Milan, Italy. Authors' Personal Archive, 2019

focal point and an important activity hub, it can become a powerful symbol and a memorable location that visitors will remember as the spirit of the city center.

5.4 Principle 3: Creating Diversity and Interest

Throughout history, the city center has always been distinguished by its diversity. Mixing uses (e.g., offices and different types of shops such as retail shops, special city centers, special markets for agricultural and weekly products, and street vending) and cultural attractions (e.g., museums, art centers, entertainments, restaurants, and residences) should be encouraged in the city center when the market is large enough to accommodate all of them (McBee et al. 1992).

In cities where, under the influence of modern urban planning, the separation of urban functions has been accepted as an indisputable principle, the single-use nature of city centers and the lack of necessary diversity in uses have rendered city centers insecure, monotonous, and boring (Khatam 2002). Mixed, vibrant, active, and dynamic uses satisfy the diverse needs of the majority of city residents. Many professionals in the urban development industry have predicted that traditional city center shopping centers will continue to be prevalent, while future models will shift toward mixed-use, urban-style projects.

As a direct result of improper planning for city center uses, the majority of city center development is compelled to violate all zoning regulations. Local governments should frequently be incentivized to promote diversity or alter the instructions and regulations so that the development of city centers can continue to be based on a variety of activities and functions (Lockwood 2003). Bentley et al. consider diversity to be one of the most important characteristics and principles of environmental responsiveness due to the significance of creating diversity. They believe that accessible locations are only valuable if they provide a variety of experiences and interests. This can be accomplished via a variety of forms, applications, and meanings. Consequently, places with a variety of uses include a variety of building types and forms and attract different people at various times and for various reasons. In addition, diversity in use is the key point in the principle of diversity as a whole (Bentley et al. 2013).

Variety and interest can respond to the need to stimulate emotions and enjoy the experience of new and pleasant spaces, or to the "quality of exploring spaces" in a nutshell. Finding and discovering are powerful motivators that cause people to enter a new space and move throughout the city in search of other spaces and events. In order to promote a sense of experience and discovery in the city center, the city center should be able to enhance the quality of diversity and interest. The creation of complexity in city center spaces (ideally, but not excessively), the occurrence of diverse events in the form of various activities, and various and attractive visual qualities in building design and streetscape can foster this sense. In this regard, Abbaszadegan (2006) states that according to the research conducted in Vancouver, Canada, the most unsuccessful spaces are those that are bare and monotonous with excessive clarity. Ordinary materials, dull colors, lack of plants and water, rough and soulless pavement, and uniform composition of space all contribute to excessive clarity, which in turn diminishes people's interest in a space because there is nothing to discover there (Abbaszadegan 2006). The architecture of buildings, the materials employed, and special events (celebrations, festivals, and funerals) can also contribute to the diversity of city centers. Given the variety of activities and uses in the city center, it should be a rich, diverse, and complex environment with many sensory stimulants. As a dynamic place, the city center should include elements that can be altered to keep it fresh, engaging, and exciting. These can include a change in the facade of a shop or a list of planned events and activities (Fig. 5.8). Although visual diversity and complexity occur naturally in many urban areas, in some urban cores they must be created. However, this diversity should not be haphazard, so as to destroy the visual cohesion of the city center.

Diversity in smaller components, including architectural details of buildings, billboards, and shop facades (which are mostly associated with sidewalks), is also important. Transparent facades on the ground floors of buildings that connect indoor activities of shops to the outdoors, the inclusion of activities that can move outdoors in

Fig. 5.8 The regular occurrence of various events in Piazza della Signoria in the historical core of Florence has had a significant impact on the variety and attractiveness of the city center, as well as the presence of people of all ages and genders. Florence, Italy. Authors' Personal Archive, 2022

good weather (such as grocery stores, cafes, and street vendors), and the planning of special events are conditions that increase variety and interest (Paumier 2004). It should be noted that the quality of diversity in the city center should correspond with legibility (principle 1). In actuality, a good city center should be "legible" without sacrificing its visual-functional attractions and complexity. In contrast, an environment that confuses and misdirects people cannot provide the basis for a quality city center.

5.5 Principle 4: Ensuring the Balance of Activities

The balance of uses, which can continually invigorate a district, is a further factor that contributes to the increase of collective interaction and vitality in the city center. For instance, large amounts of office space and its distribution in the city center can result in an empty city center after business hours. Therefore, a balanced combination and distribution of shops, visitor-attractive uses, and residential areas can contribute to the district's vitality at various times. Therefore, the high concentration of significant uses should be avoided (Paumier 2004). The urban core should be home to a variety of tenants and residents who can meet the various needs of the urban community. The city center is primarily used for shopping, entertainment, and civic purposes. In addition, hotels, residential areas, and educational institutions with dormitories are among the activities whose balanced distribution can prevent city center stagnation (Fig. 5.9) (Booth et al. 2001).

Visibility of important activity points that can attract the population is crucial to the success of a city center. Less visible parts of a site can be revitalized with the proper and balanced injection of civic uses. Today, both government officials and developers are becoming increasingly aware of the significance of designing public spaces. Public spaces are now regarded as a necessary condition for a healthy civil society. Civic uses are now viewed as lucrative factors for city centers, which can attract individuals for a variety of activities by fostering trust and diversity. Nevertheless, the incorporation of civic uses in the city center can be time-consuming. Therefore, it is recommended that city center properties be preserved for potential future uses (Bohl et al. 2002).

Overall, it should be noted that a balanced combination of activities can enhance visual values, increase security, allow people to live near their place of employment, and reduce the amount of misdirected traffic in and around the city center, thereby facilitating the presence of large crowds in city centers.

Fig. 5.9 The establishment of hotels and hostels in city centers significantly contributes to the continuation of life in these regions, particularly at night, and to the promotion of vibrancy and safety. Dijon, France. Authors' Personal Archive, 2022

5.6 Principle 5: Creating a City Center Based on Local Facilities and Resources

Geographical conditions, the process of historical development, landmark elements, and architectural heritage make city centers unique and bring about opportunities for the creation of a comprehensible, tangible structure and a unique identity (Fig. 5.10). Every community needs to know the physical conditions in the central urban area and how they can be a starting point for the design, planning, and coordination of future growth and change.

The significance of construction based on available resources and facilities is evident. However, numerous urban reconstruction projects have obliterated old contexts and replaced them with larger buildings and designs deemed more efficient. Three important conclusions can be drawn from experiences and projects (Paumier 2004):

- The human and material costs of new large-scale development are unacceptably high.
- New suburban development models that emphasize cars can disrupt the pedestrian-centered environment in the city center.

Fig. 5.10 Cities should focus on protecting and promoting their architectural heritage and local cultural identity while reaping the many benefits of globalization. This will result in city centers that are distinctive and aesthetically pleasing. Bern, Switzerland. Authors' Personal Archive, 2022

- Traditional patterns in urban cores, as well as their scale of development, architecture, and history, are valuable and significant to the members of the community.

5.7 Principle 6: Eliminating Incompatible Activities and Uses

Certain urban land uses can disturb people's comfort and leads to environmental pollution. Such pollution can be observed in various parts of the city, but the city center, which contains a dense concentration of diverse land uses, requires a greater degree of land use sensitivity than other urban areas to ensure the comfort of the crowd. There are also other uses, such as large plots (e.g., warehouses), that can have negative physical and psychological effects and contribute to insecurity in the city center.

Therefore, the exclusion of industries, manufacturing factories, and certain service centers (such as pollutant repair shops) and warehouses from the city center can improve the functional quality of these centers.

5.8 Principle 7: Ensuring Visual and Functional Continuity

A unifying visual matrix enables viewers to comprehend the urban scene as a whole and locate its intended components. It is also important for salespeople and business owners because it makes their stores and shops easily identifiable to customers, thereby ensuring their success and profitability. Visual and functional continuity will be easier to achieve if a strong organizing structure with a regular pattern of

Fig. 5.11 The integrated and unified visual structure encourages pedestrian movement and directs people to the center while promoting public uses. Rome, Italy. Authors' Personal Archive, 2019

passages, equal-sized blocks, visible spaces at the right locations, and good connections between buildings and streets are established. Coherence and continuity also rely on the precise operation of the city's other design elements (Paumier 2004) (Fig. 5.11).

The experiences of the 1970s and 1980s demonstrate that the introverted focus of projects and methods that separate people from the streets, such as the use of overpasses and underpasses, can result in the demise of city centers. Boston's Lafayette Place (a commercial complex with shops, hotels, etc.) is one example. The street should be impressed with the authority and credibility of the city center's significant structures. Instead, the Lafayette Place building "turns its back to the street, making it appear more like a prison than a row of storefronts. Its empty walls and locked doors insult the city and its inhabitants. Lafayette Place is a suburban mall dropped into the middle of a city, an inward-looking building with all its life hidden inside, and nothing but blank walls facing toward the world around it" (Campbell 1986).

Rather than encouraging visitors to see and experience other buildings and spaces in the city center or nearby neighborhoods, such projects seek to trap visitors (McBee et al. 1992). Therefore, such projects in the city center should be avoided, unless such collections include a public space with a specific use in relation to the street axis, rather than an inward-looking focus (LMN Architects 2005b).

5.8.1 Architecture

Although equalizing the height of buildings can be an important factor in establishing a sense of visual continuity, it cannot be definitive or absolute due to the visual error that is inherent to human vision. However, precise transitions between buildings of varying heights can aid in minimizing discontinuities and enhancing cohesion. Due to its greater impact on the pedestrian experience, continuity on the ground floor of buildings is of greater significance (Fig. 5.12). Here, the use of similar building materials and the proportion and form of buildings should be encouraged. Similarly, the distance between buildings and a positive relationship between their locations should be considered, as well as the ratio of openings in building facades and the

Fig. 5.12 The greater visibility of the ground floor layer and the concentration of key and appealing uses in this layer, as opposed to other parts of the building, results in enhanced perception, enhanced mental durability (stronger lasting images in the memory), and a positive effect on the pedestrian experience. Bologna, Italy. Authors' Personal Archive, 2019

organization and creation of high-definition defined views from the ground and upper floors. In actuality, the objective is to coordinate the design of city center buildings and transform the individual facades of each block into a macro-architectural unit.

Though continuity needs compatibility, it also allows for variety in design. A basic solution is to recognize specific methods for using design elements to create an architectural combination and repeat selected design details from one building to another. The more the buildings take on a cohesive and centralized form, the more important their continuity will become. Hence, in smaller city center areas with fewer buildings, integration is more important and contrast in forms, materials, volumes, and height of buildings is less accepted (McBee et al. 1992).

5.8.2 Streetscape

Well-designed streetscapes can play a significant role in establishing a sense of identity for the central urban area, as well as in creating visual cohesion and enhancing the physical structure's clarity. A hierarchy of street landscape designs must be used to visually improve the functional hierarchy of streets. The intermittent use of essential street furniture such as lighting fixtures, pavements, benches, trees and plants, press kiosks, and public phone booths creates a visual overlay that reinforces the streets-and-blocks-based organizing structure of the city center.

5.9 Principle 8: Maintaining Relative Enclosure with Human Scale

Tavassoli (1997a) identifies enclosing space as the first principle governing the design of urban spaces, stating, "If space is not enclosed in an aesthetically pleasing manner, an attractive urban place cannot be created." This principle is common in the old cities of most countries in the world. The spaces between buildings must be designed in a way that evokes human emotion. Any space, regardless of its size, height, width, or depth, can be visually characterized to evoke reactions and emotions. According to this thinking, "the human is the most important element of urban space, so it should be designed on a human scale" (Tavasoli 1997a). Large-scale new projects should not have a negative effect on the human-centered spirit of city centers, despite the importance of density in creating concentrated urban masses. In fact, the scale of buildings and passageways can alter the sensory experience of pedestrians in the city center and, consequently, the activities that take place there. The scale of construction can vary between small and large urban cores. Nevertheless, special design measures (such as transitions, setbacks, buffers) should be taken to reduce the negative effects of such construction if it is necessary to create density and increase building floors in the cores of larger cities or the major city centers (Fig. 5.13).

Fig. 5.13 It is necessary to consider human scale when designing city center structures. In this area of the city, the required density does not necessitate that building sizes and heights vary in an unusual manner. If it is necessary to construct a large structure, the negative impact should be mitigated through design (such as articulation and volume fracture, setback). Bologna, Italy. Authors' Personal Archive, 2019

5.10 Principle 9: Maintaining City Centers' Concentration and Compactness

The best and most successful city centers in terms of physical characteristics are those that are compact and centralized, as they facilitate access and maximize convenience for users, thereby facilitating economic interaction between uses. In many cities around the world, the growth and expansion of city centers have created large gaps between buildings and uses that must be adjacent to one another for maximum efficiency (McBee et al. 1992). This widespread and unrestricted growth of city centers can discourage pedestrian movement and reduce connections between city center activities and uses, resulting in a decline in the economic and social vitality of these centers.

5.11 Principle 10: Increasing Accessibility and Pedestrian and Vehicle Access to the City Center

Accessibility is an advantage of cities over other human settlements. In the absence of proper access, the potential for utilizing spaces is diminished, as is the sense of community. According to Lynch, access can be classified based on the elements to be accessed and the entity paying for them. Access to people, acquaintances, etc. is possibly the most fundamental type. Access to certain human activities, specific goods and resources, places (such as open spaces, civic centers, and the natural environment), and access to information follow (Lynch 1984). In the book Responsive Environments, access is defined in terms of "physical permeability" and it is hypothesized that the permeability of any system of public spaces is dependent on the number of potential routes provided from one point to another and that only accessible spaces can give people the ability to choose. Consequently, access is regarded as a crucial factor in evaluating the responsiveness of an environment (Bentley et al. 2013).

Francis (1989) describes three types of access: physical, social, and visual. Carr et al. (1992) enhance this classification with symbolic access.[1] Physical access, which is closely related to social access, is crucial for our current focus and can significantly improve the quality of city centers.

Important factors in the formation of the city center as a place are those that facilitate movement among the various activities and uses of the city center. Easy

---

[1] Types of access include:

(A) Physical access to urban spaces: moving from one point to reach other points in the city.

(B) Social access: the possibility of all social classes using and enjoying the space

(C) Visual access (or visual permeability: the possibility of being seen in the space).

(D) Symbolic access: the presence of symbols in the space indicating people whose presence in the space is deemed pleasant or unpleasant. For example, some uses and functions (e.g., a special category of shops, playgrounds of different ages, etc.) encourage a specific group of people to enter the space.

and well-defined connections between offices and residences, as well as commercial, recreational, and cultural uses, increase pedestrian activity and help city center activities support one another. Contrary to appearances, the global experiences of the last few decades indicate that closing city centers to cars in the name of pedestrianization can have a negative impact on the life and continuity of life in such centers (Paumier 2004). Therefore, measures should be taken to facilitate vehicular and pedestrian access to the city center. In fact, easy and convenient automobile access and the availability of adequate parking lots are also essential to the success of the city center as a market. Nonetheless, a planning method that always prioritizes the flow of automobile traffic and encourages the creation of low-cost parking lots for personal development substantially degrades pedestrian environments. Every street in the city center should be designed with pedestrians in mind. Especially on retail streets, car traffic and required parking spaces should not be neglected and should be carefully designed so as not to disturb the human scale of the streets.

Parking lots, large multi-story parking structures, and very wide streets should not create a gap or barrier between activity axes, as this creates an unsafe and unfavorable pedestrian environment. Consequently, the sections that follow concentrate on pedestrian movement, the creation of parking lots (the primary obstacle to pedestrian access to the city center), and the use of public transportation in city centers (Fig. 5.14).

5.11.1 Pedestrian Movement

A priority in creating the physical context (as a factor that stimulates regeneration) is making the city center a comfortable place for pedestrians. Most cities should have central areas with activities placed within 10- to 15-min walking distance (Paumier 2004).

This helps the city center overcome one of its obstacles, which is to establish a beautiful and dynamic city center where people can easily drive, park, and stroll to various locations. A dense area, a clear structure, and an integrated sidewalk linking system are needed to meet this challenge. In addition, short blocks and access roads between blocks can make it easier to move through streets. Defined connections among parking lots and shopping malls are also essential.

5.11.2 Parking

City center parking can include on-street parking, private and public parking lots, structured parking, and garages. Undoubtedly, the provision of parking space is an overlooked yet essential element of city center design.

Parking is often cited as the cause of many economic problems in the city center if they cannot play their role well (McBee et al. 1992). To increase the ease of use of

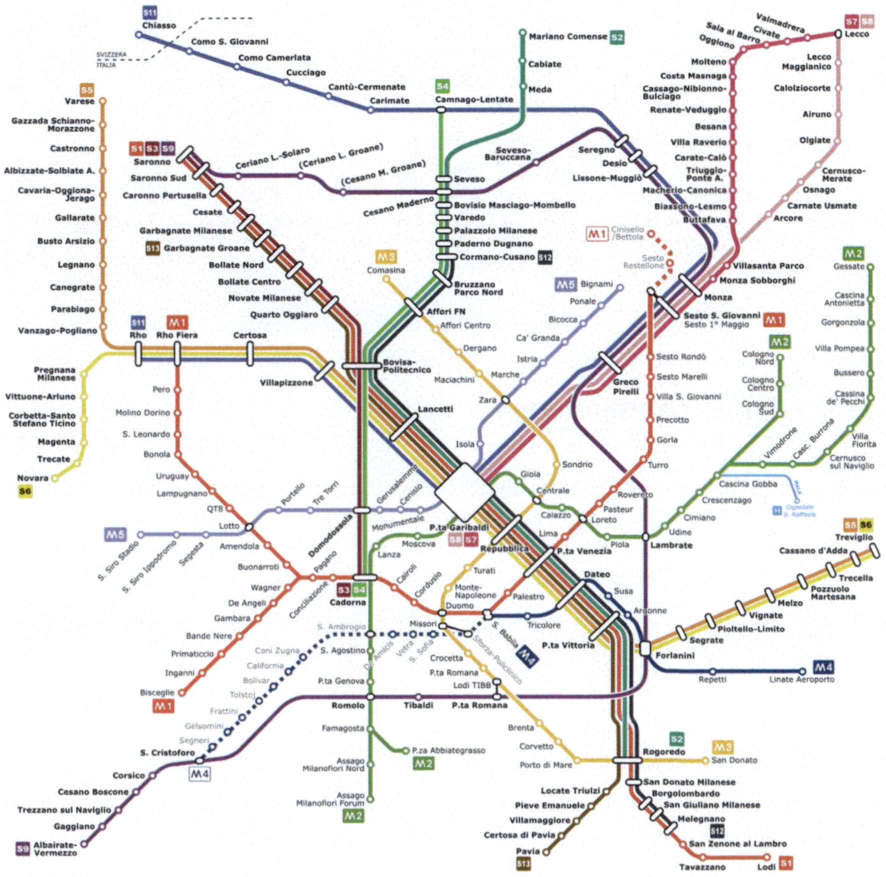

Fig. 5.14 Map of Milan Suburban Railway and Metro Services. This makes it possible to use public transportation to reach all parts of the city, including the city center. Milan, Italy. *Source* Friedrichstrasse (2012)[2]

the city center and its attractiveness as a shopping and entertainment center, parking-related policies should give priority to customers and visitors over employees who need to use the parking space on a daily basis. Signs that help buyers find parking spaces as well as plans that reduce the costs for users can increase the power of the central area in competition with suburban shopping malls. Of course, there should be parking for the long-term use of staff. This parking should be located farther from the staff's workplace, within 4 to 5 blocks of walking distance, or further away in locations connected to the city center by taxis or buses. Fixed or variable price rates, based on short-term or long-term needs for space, place and time of day, etc. are among the key factors in the effective planning of parking) (Paumier 2004).

[2] https://commons.wikimedia.org/wiki/File:Milano_-_mappa_servizio_ferroviario_suburbano_e_metropolitana_%28schematica%29.svg.

Fig. 5.15 The vast majority of Florence's historical fabric contains bus tracks. This transportation system with a very regular schedule increases pedestrian access to the city's historical core while preserving the tranquility of the city center. Florentine Italy. Authors' Personal Archive, 2019

5.11.3 Public Transportation

A public transport or loop system can be used to connect the main centers of activity within the city center and to link the city center to the suburb, acting as an alternative to cars in the center. To encourage people to use public transport, time intervals between consecutive public transport vehicles should be short and the fares should be reduced as much as possible. Providing the required parking with affordable prices can also be considered to prevent car travel within the city center (Fig. 5.15).

5.12 Principle 11: Security and Safety of City Centers

To be considered a place for social interaction and public space, the city center must make people feel secure and comfortable. In actuality, a comprehensive revitalization of the city center is only possible under certain conditions in which the environment is safe and secure (Oc and Tiesdell 1998). Given the concentration of resources and financial, administrative, commercial, and service facilities in the city center, the population density is high during the day but drops significantly at night and on weekends and holidays. In other words, the significant difference between daytime and nighttime population densities is a significant problem in urban cores, resulting in issues such as a diminished sense of social security, poor spatial vitality. In fact, the greater the population in the city center, the more social security for citizens and space users (Kokabi 2005). Also, the design of different parts of the city center, such as the between-block paths, routes to the main parking lots, passages to long empty walls with no opening, and open spaces inside or connected to the city center (such as plazas, parks, green spaces), can be implemented in such a way as to reduce the insecure spaces without urban defense, thereby playing a crucial role in enhancing this aspect of the design. The streets should be visible from adjacent buildings, and public spaces should provide unobstructed views. In fact, only spaces that lack hidden, secluded, and crime-provoking points (defenseless and non-dynamic spaces) can be utilized by individuals of all ages and genders. However, there are a number

Fig. 5.16 The lighting of alleys, the emphasis on keeping a portion of shop window lights on during closing hours, and the establishment of residential uses with maximum openings to the thoroughfare increase the psychological security of pedestrians in city centers. Florence, Italy. Authors' Personal Archive, 2022

of other measures, such as indirect supervision of residents from residential units in the city center, physical restoration of city centers, the development of shopping centers, an increase in mixed uses, the development of dynamic open and recreation spaces, the establishment of police stations, that can improve the quality of life in city centers by providing residents and users with a sense of security (Fig. 5.16).

On the other hand, the safety of people in the city's center must also be ensured by implementing the appropriate design measures. Moreover, a good urban core should be physically and environmentally safe. The ability to move safely in cold seasons (despite snow and ice on the roads), the decreased risk of collision between vehicles and pedestrians, etc., can enhance the safety of city centers (Fig. 5.17).

5.13 Principle 12: Comfort and Well-Being

Ensuring the physical and mental well-being of pedestrians is very important. A comfortable and safe environment for pedestrians requires adequate sidewalk width, shade, seating, and a sense of protection from car traffic. Activity, visibility, and a constant eye on the street are essential to creating security. The five important factors involved in planning for physical comfort, physical safety, and maintenance are weather, traffic, amenities, physical safety, and maintenance.

5.13.1 Weather

Since weather plays an important role in the comfort of pedestrians, tall buildings should be located in such a way that their shadows do not cover important public places, especially in the cold months of the year. Buildings located along east–west streets limit sunlight to the south side of sidewalks, while streets located north–south

Fig. 5.17 The installation of fire hydrants in all passageways of historic districts improves the safety of city centers. Venice, Italy. Authors' Personal Archive, 2022

enjoy more sunlight, which increases the activity of pedestrians in the important axes located in this direction. In hot weather, trees or shop canopies provide shade (Fig. 5.18). It is also important to include indoor public areas to protect against cold and stormy weather. The use of upper-floor setbacks in buildings with a height of more than 10 floors, which can reduce the effects of air tunnels on streets and public places, should be encouraged among builders and owners and controlled via construction codes (Paumier 2004).

5.13.2 Traffic

Planning for street design should be able to create a sense of separation and protection for pedestrian areas against vehicle lanes. The volume and speed of vehicles on streets with a high volume of pedestrians should be kept at an optimal level so as to reduce noise and pollution, increase the safety of pedestrians, and on the other hand, prevent a reduction in the dynamics of the city center that may result from banning cars (Fig. 5.19).

Fig. 5.18 The use of opaque canopies with the possibility of passing sunlight moderates the intensity of light and creates suitable shade in the passages and creates a suitable shelter for pedestrians during rains. Geneva, Switzerland. Authors' Personal Archive, 2022

Fig. 5.19 The complete prohibition of cars in certain areas of city centers should be implemented very intelligently; in such a way that, while maintaining the comfort of pedestrians and creating a cozy and safe place, it allows for their access by car in the vicinity of the mentioned spaces. Piazza del Popolo, Rome, Italy. Authors' Personal Archive, 2019

Fig. 5.20 Temporary shaded pavilions for pedestrian resting near the Geneva Lake, Geneva, Switzerland. Authors' Personal Archive, 2022

5.13.3 *Amenities*

Seating areas should be located in public areas and at bus stops for the convenience of those utilizing public transportation. In order to accommodate the large number of pedestrians, sidewalks must be sufficiently wide to accommodate stalls, bus stops, street trees, and street furniture (Fig. 5.20).

5.13.4 *Physical Safety*

Principle 11 emphasized the safety of urban centers. One can also examine safety from the perspective of creating comfort. Materials used for paving should provide safe surfaces and avoid unnecessary alterations that pose a risk to pedestrians. Visible crosswalk lines are necessary to promote a sense of safety.

5.13.5 *Maintenance and Preservation*

The upkeep of roads and public spaces, as well as the cleanliness of the city center and the maintenance of its organization, can increase the physical and mental comfort of individuals (Fig. 5.21).

Fig. 5.21 A concrete example of maintaining and preserving the city center cores is paying attention to the necessity of renovating the city center's pavements, particularly during the days when the spread of the Coronavirus drastically reduced the number of tourists in these areas. Florence, Italy. Authors' Personal Archive, 2022

5.14 Principle 13: Design Based on Different Tendencies and Tastes

In the current conditions, urban planning and development approaches that are based on the tendencies of urban dwellers have become more popular. It should be noted that this stems from people's desire for places where they can communicate with each other (Mattson-Teig 2004). Obviously, creating a place, or creating vibrant pedestrian-centered spaces where different people with different motivations, tastes, and inclinations are able to communicate is essential to the sustainable development of the city center (Schmitz et al. 2006).

Consumers now view shopping as a means of escape, as well as a recreational and social activity because it requires interaction with others. This emphasis on creating experiences for buyers is explained in the Experience Economy by Pine and Gilmore. They explain how shops, restaurants, and entertainment venues now prioritize the design and decoration of their establishments over the quality of their services and enhancements. In this manner, they appear more valuable, resulting in increased prices and profits (Fig. 5.22). (Pine et al. 1999). Therefore, creating urban places is less directly focused on sales and more focused on providing the consumer with an experience. The city center is intended to have a memorable quality by incorporating design elements that influence how consumers perceive their experiences (Beyard 2003).

To perceive a positive meaning from space, it must be consistent with how people live and encourage a pattern of activities that facilitate communication with space.

Fig. 5.22 Some individuals find entertainment and pleasure in looking at storefront windows. This encourages shop owners in city centers to invest in aesthetically pleasing and distinctive window displays that appeal to their target demographics. Milan, Italy. Authors' Personal Archive, 2019

Planning common objectives for space users enhances their connection to space. The space will respond to the needs and rights of users and enrich their lives in this instance. If the space operates at a higher semantic level, it will be stimulating enough to evoke the memories and experiences of individuals, families, groups, and cultures. In this instance, each individual is connected to a larger group, a shared memory, or an experience. Consequently, the environment fosters a sense of significance, affiliation, and concern for the lives and destinies of others (Carr et al. 1992).

Creating a place or vibrant pedestrian-centered spaces where people can communicate sensually and experientially is unquestionably essential to the continued growth of the city center (Schmitz et al. 2006). To create sensory and experiential desirability in the city center, one must consider all of these groups simultaneously, given the diversity of groups exploiting city centers. Attention to different age groups (such as children, adolescents, and the elderly), gender groups (women and men), psychologically distinct groups (active and passive people in the environment),[3] and groups with special needs (disabled, blind, etc.) in the city center requires a thorough understanding of these groups and a strong design ability that is tailored to their needs. Positive responses to such tendencies include the development and creation of recreational and sports spaces for children and adolescents (e.g., skating rinks), encouraging and providing spaces for activities of indigenous people (e.g., street performances), and maintaining a strong connection with the city center (Fig. 5.23).

[3] Abbaszadegan quotes White as stating: "Passive or inactive presence in space is a state in which one observes the environment but does not speak or do anything specific, but just stands and watches people. The presence of certain physical elements in the city can increase passive presence in space. Active presence in space requires the creation of an environment where people who need more social interaction can find a space to participate in various activities" (Abbaszadegan 2006).

Fig. 5.23 In addition to physical-spatial design changes, diverse and appealing events should be organized in urban centers in order to make them appealing to a variety of demographics, including children. Bologna, Italy. Authors' Personal Archive, 2019

5.15 Principle 14: Emphasis on High Quality

The simplicity of design, the use of high-quality materials, and the maintenance of public spaces have a direct impact on the quality of pedestrian experiences and encourage the use of public areas. The materials used for street landscaping and street furniture should be of the highest possible quality. Priority should be given to the primary components of a high-quality public environment, such as sidewalks, trees, lighting, and seating.

Design measures should take into account a minimum requirement for long-term maintenance costs, which will be covered by the Initial Capital Funding Strategy. The creation of design guidelines and the subsequent review of designs can enhance the quality of private and public sector developments (Paumier 2004).

5.16 Principle 15: Balance Between the Static and Dynamic Aspects of Pedestrian Movement

Spatial visualization in the city center results from corridors and nodes or different combinations of the two. However, the need to maintain a balance between the static and dynamic movement of pedestrians in urban centers is of utmost importance. Therefore, providing facilities and amenities to create quiescence and using activities, functions, as well as environmental qualities and street landscape to encourage stops in city centers as part of a planned pedestrian movement should be considered in the city center design. Stopping in front of store windows is one way that pedestrians can slow down. Nonetheless, since there is not always a desire for such stillness, other arrangements can be made to this end, including but not limited to street performances and competitions, creating plazas and fronts, parks, and playgrounds for children, providing proper furniture for relaxing close contact with city centers and creating street art galleries. Likewise, the use of special flooring, less dynamic facades, fewer spatial enclosures, appropriate urban furniture, space openings (in front of government and public spaces) and at the entrances to city centers,

Fig. 5.24 The ability of outdoor cafes to induce a pause in the behavior of pedestrians, while fostering communication between the inside and outside users, contributes significantly to the vitality of city centers, Piazza della Signoria in the historical core of Florence. Florence, Italy. Authors' Personal Archive, 2019

and providing conditions for some outdoor activities such as cafes or street restaurants that can encourage people to stop are factors that can provide the ground for creating the static conditions required by pedestrians (Fig. 5.24).

References

Abbaszadegan M, Rezazadeh R (2002) Looking at the plan to preserve and restore the historical context of Semnan. Haft Shahr J 2(4)

Abbaszadegan M (2006) Social-psychological dimensions of urban spaces. Int J Eng Sci 16(1)

Bentley I, McGlynn S, Smith G, Alcock A (2013) Responsive environments. Routledge

Beyard MD (2003) Ten principles for rebuilding neighborhood retail. Urban Land Institute

Bohl CC, Schwanke D (2002) Place making: developing town centers, main streets, and urban villages. Urban Land Institute

Booth G, Urban Land Institute (2001) Transforming suburban business districts. Urban Land Institute

Campbell R (1986) The very complex Lafayette Place. Lafayette Place Associates

Carr S, Francis M, Rivlin LG, Stone AM (1992) Public space. Cambridge University Press

Francis M (1989) Control as a dimension of public space quality. In: Public places and spaces, human behavior and environment: advanced in theory and research; V.10. Retrieved from http://online.sagepub.com

Friedrichstrasse (2012) Schematic Map of Milano Suburban Railway and Metro Service [Map]. Wikimedia Commons. Retrieved from: https://commons.wikimedia.org/wiki/File:Milano_-_mappa_servizio_ferroviario_suburbano_e_metropolitana_%28schematica%29.svg

Khatam A (2002) Massing of housing in the center of the city. J Mass Builders Hous 3 and 4

Kokabi A (2005) Planning the quality of life in urban centers, definitions and indicators (master's thesis). Tarbiat Modares University

LMN Architects (2005b) Mill Creek Town Center Design Guidelines. Mill Creek, Washington, DC

Lockwood C (2003) Raising the bar: town centers are outperforming traditional suburban real estate products. Urban Land 62(1)

Lynch K (1964) The image of the city. MIT Press

Lynch K (1984) Good city form. MIT Press

Mattson-Teig B (2004) Lifestyle retail: a new generation of retail centers is aimed at meeting changing consumer and community needs. Urban Land 63(2)

McBee S, Urban Land Institute, Basile RJ, Urban Land Institute Urban Development/Mixed-Use Council, Petersen DC, Phillips PL (1992) Downtown development handbook. ULI—The Urban Land Institute, Washington, DC

Oc T, Tiesdell S (1998) City centre management and safer city centres: approaches in Coventry and Nottingham. Cities 15(2):85–103. https://doi.org/10.1016/S0264-2751(98)00005-3

Paumier C (2004) Creating a vibrant city center: urban design and regeneration principles. Urban Land Institute

Pihlak M (1988) The city of the 21st century. Arizona State University

Pine BJ, Gilmore JH (1999) The experience economy: work is theatre & every business a stage. Harvard Business School Press

Schmitz A, Scully J, Urban Land Institute (2006) Creating walkable places: compact mixed-use solutions. Urban Land Institute

Tavasoli M (1997a) Principles and methods of urban design and residential spaces in Iran. Iran Urban Planning and Architecture Studies and Research Center Press, Tehran

Chapter 6
Design Strategies for Dynamic and High-Quality City Centers

6.1 Introduction

The preceding chapter outlined fifteen principles for urban revitalization and quality enhancement. This chapter discusses strategies for street landscape design, the architecture of city center buildings, and the traffic system (with a focus on parking lots) in order to provide the necessary theoretical foundation for extracting design guidelines in urban centers (in the next chapter). Thus, at the beginning of each section, the effects of such strategies on achieving the fifteen principles outlined in Chap. 5 are discussed so that, through a systematic approach, the reader is well-informed of the correct sequence of content, the relationship between theoretical context and the fifteen principles, and the strategies necessary to realize these principles.

This chapter aims to familiarize readers with city center problem-solving strategies in greater depth so that in the book's final chapter, they can understand how to use urban design guidelines to address and solve these issues.

6.2 Street Landscaping Strategies

6.2.1 Which Principles of Designing Successful City Centers Are Met by These Strategies?

Using street landscaping strategies in the city center can help achieve the following principles from the fifteen principles proposed in the previous chapter:

Principle 1: Development based on an organizing pattern
Principle 2: Formation of a distinguished identity
Principle 3: Creating diversity and interest
Principle 5: Creating a city center based on local facilities and resources

A. Shakibamanesh and M. Ghorbanian, *Designing Responsive City Centers*, Design Science and Innovation, https://doi.org/10.1007/978-981-97-8911-5_6

Principle 7: Ensuring visual and functional continuity
Principle 8: Maintaining relative enclosure with human scale
Principle 11: Security and safety of city centers
Principle 12: Comfort and well-being
Principle 13: Design based on different tendencies and tastes
Principle 14: Emphasis on high quality
Principle 15: Balance between the static and dynamic aspects of pedestrian movement.

6.2.2 Application of Landscaping Strategies

The importance of proper street landscaping in establishing a sense of identity, a distinct physical structure, and visual cohesion in the city center cannot be overstated. A collection of street landscape designs should also be used to visually enhance the functional hierarchy of streets. The intermittent use of essential street furniture such as lighting fixtures, pavements, benches, trees and plants, newsstands, and public telephone booths creates a visual overlap that reinforces the streets and blocks that make up the urban core's organizational structure (Fig. 6.1).

Continuity of the streetscape can be achieved through urban design rules and guidelines that emphasize the harmonious enhancement of the city's streetscapes at no additional cost. However, in some areas, short-term planning can also be used to improve the streetscape by means of private renovations and new construction (McBee et al. 1992).

Using the streetscape to create an attractive and comfortable environment for pedestrians, allocating spaces to pedestrians and cars with equal priority, and establishing a positive relationship between the street and the buildings that define its edge are important design considerations for the pedestrian network system in the city center (Paumier 2004). The street landscape design in city centers should provide a unified image and a defined visual structure for the center, as well as an inviting and comfortable environment for pedestrians. Simplicity and consistency are the pillars of successful design. Simple design concepts executed with quality materials

Fig. 6.1 Integrated structure of the urban landscape in the streets of the city center can significantly contribute to the development of a sense of identity and organizational structure. Milan, Italy. Authors' Personal Archive, 2019

are the most durable over time, both in terms of upkeep and aesthetic and visual value. The landscape architecture of a street should emphasize the linear continuity of street space and increase its adaptability to accommodate a variety of uses. It should create visually appealing backgrounds for commercial activities and provide a suitable environment for other activities in the city center, without diminishing the visual appeal and visibility of street views.

(A) Street Furniture

The success and effectiveness of urban spaces are directly related to the quality of their spaces, which is the result of amenities and equipment installed in conformance with their roles and functions. This is why the urban furniture issue is so crucial to the design process (Bahreini 2020).

Utilizing well-designed furniture throughout the city contributes to its cohesiveness. There is a need for well-considered criteria for the selection and use of streetlights, chairs, trash cans, portable plants, bus stops, tree preservation tools, and vendor carts in the reconstruction of city centers. These criteria should also govern the design and placement of signposts in order to minimize their visual impact and enhance their legibility. In addition, it is necessary to establish rules and criteria for the placement and use of public artworks such as sculptures, murals, and banners. In addition, pedestrian scale lights with a height of 3.7 m should be installed to create an area with adequate services and amenities along sidewalks. These human-scale short lights can be used between intersections illuminated by standard streetlights to create uniform lighting without harsh and glaring effects and to increase the safety of passageways (LMN Architects 2005b) (Fig. 6.2).

Well-designed seats distributed at appropriate intervals are important for increasing the comfort of pedestrians, but sidewalks are not always a good place for their installation. With the exception of bus stops and cafes that place their furniture outside the shops, areas that are away from the street, but connected to sidewalks are better places for seating. As part of the design of buildings, palaces, fronts, and parks, such seating areas can be created using edges, steps, low walls, removable tables and chairs, and common benches. They can be designed on street sides where the equipment and amenities of the sidewalk allow for enough space to arrange benches perpendicular to the street. This layout creates opportunities for people to watch. Benches that face the street rather than the sidewalk are only appropriate for taxi and bus stops. Simple wooden or steel chairs are also good options for seating because of their strength and comfort. However, if there is not enough budget to buy quality benches, it is better to avoid using them altogether (Paumier 2004) (Fig. 6.3).

(B) Width of Sidewalks

The appropriate width of the sidewalk in the pathways that form the backbone of the city center and the main streets is 6 m, of which 3.6 m is allocated to the sidewalk and the area connected to the shop fronts, allowing easy passage of the passersby and providing enough space for people who stop to watch the shops. The area connected to the curb or the street edge shall be 2 m wide, while 6 m provide enough space for furniture, street cafes, and public arts used in street landscaping without disturbing

Fig. 6.2 The human scale of the lights installed in pedestrian pathways, along with good lighting, their proper placement, and the correct distance between each other, are required to bring nighttime activity to the city center. Dijon, France. Authors' Personal Archive, 2022

the movement of pedestrians. Streets intended for high volume of car traffic need an amenities area with a width of 4.6 m in addition to the mentioned width to be able to place stations and waiting points on sidewalks. In places with a lower potential presence of pedestrians, for example on sidewalks or streets in smaller cities, sidewalks with a width of more than 4.3 to 4.9 m can weaken the sense of vitality and activity in the city center (Paumier 2004) (Fig. 6.4).

(C) **Pavement of Sidewalks**

Specialized sidewalk paving on the principal city center pathways (as well as the main streets) has a significant impact on fostering a sense of comfort and visual richness. When designed and utilized in an integrated fashion, these pavements can serve as a visual link to strengthen the pedestrian system. The initial implementation of these pavements is more expensive than that of concrete, but their durability and impact justify the cost. However, the first rule for all types of sidewalks should not be forgotten: provide opportunities for walking in all weather conditions for people of all ages and footwear types. Uneven paving and shallow edges and steps can endanger pedestrians' safety and discourage their activity (Fig. 6.5).

The same paving materials are utilized throughout the network of sidewalks. To ensure the complete construction of concrete sidewalks, such paving should be used across the entire width of the sidewalk and from storefronts to the curb, as well as in amenity areas adjacent to curbs (bus stations). All sidewalks should be paved with cement pavement. Drivers can also be alerted to pedestrian crossings through the use of distinct pavements. In cold climates, snow removal equipment and machines

Fig. 6.3 The design and arrangement of seats in relation to one another, the axis of movement, and existing uses are essential for their optimal performance in the city center. Figure top; Dijon, France. Figure bottom; Lyon, France. Authors' Personal Archive, 2022

Fig. 6.4 Depending on the significance of the axis, the street's uses, and the number of pedestrians, it may be possible to consider a wider sidewalk than the standard minimum width. Avenue Champs Elysees, Paris, France. Authors' Personal Archive, 2022

Fig. 6.5 In addition to facilitating the movement of pedestrians, the pavements used in passages contribute to the integration and connection of key urban spatial structures. Lyon. France. Authors' Personal Archive, 2022

can damage fragmented street intersection flooring. The most effective pedestrian crossings in Europe feature embossed zebra patterns at street level to draw drivers' attention.

(D) **Street Planting**

Throughout Iranian history, trees and green spaces have been utilized for both religious and climatic reasons. The sanctity of water and plants because of their association with paradise, on the one hand, and the hot and dry climate of the country, on the other, has prompted the use of plants as a basic natural need on various scales, including yards, gardens, shrines, and promenades (Bahreini 2012). Today, one of the most important goals of urban designers is to improve the quality of urban environments. In today's world, to be livable, cities must meet the requirements of a natural ecosystem in terms of humidity, temperature, light, air composition, and exchange. Additionally, plantations can enhance the physical, spiritual, and psychological aspects of urban areas. In this regard, it is necessary for urban areas to achieve a reasonable balance between built and natural spaces.

In fact, street tree planting is an investment to improve the quality of a city's public spaces. Where upper-floor architecture lacks continuity and inspiration, tree canopies can provide a unified visual structure without obstructing the view of stores, billboards, and signs from sidewalks and streets. To improve the linearity of urban street space, trees should be planted at equal distances from the street curbs at regular intervals.

Care should be taken in selecting the types of trees, the amount of sunlight required, the size of the tree umbels, and their deciduous nature in the form of informed planning. Planting umbrella trees in the pathways is a crucial element of the streetscape in the city center because they create an integrated and high-quality foreground from the driver's perspective and create a sense of separation between the sidewalks and the automobile traffic. In addition, their shading can help soften the air and create comfortable conditions. Indicators of human scale, and trees mitigate the effect of tall buildings and calm pedestrians without obstructing storefront visibility (Fig. 6.6).

Although raised, fixed flower boxes and planters are used to increase the sense of separation between sidewalks and streets, they may limit the space for pedestrians and reduce the space's versatility, particularly in amenity areas. In addition, they can create a sense of disorder and confusion among pedestrians and increase the cost of constructing and maintaining the streetscape. Due to these factors, their use is discouraged unless they are the only option for creating green surfaces and plantations on non-flat surfaces and installation lines (Shakibamanesh 2008). In addition, if there is a tendency to use colorful flower arrangements as part of the street landscaping,

Fig. 6.6 Proper use of water and plants helps to soften the artificial atmosphere of city center areas, while also providing pedestrians with a place to rest and enhancing a sense of vitality. Dijon, France. Authors' Personal Archive, 2022

Fig. 6.7 Most beautiful definition of the boundary between the street and the sidewalk is provided by the colorful trees. La Spezia, Italy. Authors' Personal Archive, 2019

flowers, and portable plants may be used at regular intervals and within the boundaries of the curbs. However, a sufficient annual budget will be required for the seasonal planting and maintenance of such plants (Fig. 6.7).

(E) **Fountains**

Fountains play an important role in enhancing the streetscape, particularly in areas where the structure of the transportation network has been altered and where redevelopment plans have left vacant lands and buildings. When fountain areas are reconstructed with new attractions, they alter the city's landscape and play a significant role in establishing a new identity (Fig. 6.8).

(F) **Public Arts**

A strong emphasis on art, including the use of sculptures, fountains, and graphics, can be a defining characteristic of the city center. Other components, such as tree networks and guards, benches, manhole covers, fences, and signs, can also be meticulously designed to create artistic elements that are remarkable and admirable. In addition to protecting the identity of the city center, public art can help humanize the environment. It can be humorous or emphasize the historical significance of a location to attract people's attention and give their urban experience meaning. When art is incorporated into the overall design of a public space, plaza, park, or sidewalk, it will be more successful (Fig. 6.9).

(G) **Public Open Spaces**

When public open spaces are thoughtfully incorporated into the overall pattern of streets and blocks, they improve the legibility of the city center by creating highly visible landmarks. These include linear open spaces or boulevards that help define the primary axes of vehicular and pedestrian movement, green spaces that mark the entrances and exits to city centers, and public parks and gardens that define activity centers (Fig. 6.10). These open spaces create a common focal point upon which the

Fig. 6.8 City centers can utilize water fountains on a variety of scales. The above image depicts a local fountain in the historical fabric of Florence, Italy. In the picture below, an iconic fountain in the form of a spectacular work of art in the center of Lyon, France can be seen. Authors' Personal Archive, 2022

Fig. 6.9 Presence of various forms of public art has transformed the historical center of Florence into an alluring museum for tourists and locals alike. In these passageways, the existence of some valuable works of art, such as pastel paintings on the floor, is limited to one or two days; consequently, they can be considered a form of temporary public art. Florence, Italy. Authors' Personal Archive, 2022

Fig. 6.10 Location, type of spatial distribution, and proximity of Milan's central park to the city's historical core have transformed it into a beautiful entrance and inner courtyard for the center of this city. This park's high level of activity for various age and gender groups has contributed significantly to its use and presence of residents and tourists in the city's core. Milan. Italy. Authors' Personal Archive, 2019

development and establishment of different urban land uses can be based. Such spaces serve as anchors that place small developments within the context of the city center, thereby creating a cohesive, aesthetically pleasing whole (Paumier 2004). There is a need to fund the creation of such public open spaces in numerous urban cores where they do not exist. Despite the fact that the size of public open spaces is not a primary criterion, they should be designed to maximize accessibility and placed in the city's visible center in order to function more effectively as an element of identity. These spaces must be adaptable to a wide variety of functions and connect the boundaries of planned activities. In addition, they should provide a specific and varied number of seating areas and use the highest quality design and materials possible.

Local parks and smaller public gardens may also be accredited components for city center subareas. These open areas must be a part of an integrated sidewalk system and create a positive visual structure for the central area as well as a distinct local identity for their neighboring structures (Fig. 6.11).

6.3 Architectural Solutions Regarding City Center Buildings

6.3.1 Which Principles of Designing Successful City Centers Are Met by These Strategies?

Using architectural solutions for city center buildings can help achieve the following principles from the 15 principles proposed in the previous chapter:

Principle 1: Development based on an organizing pattern
Principle 2: Formation of a distinguished identity

Fig. 6.11 Presence of local parks at the intersections of the city center's neighborhoods can help to maintain a sense of life throughout the majority of the day and night in these areas. Locating uses such as outdoor cafes, schools, and public and religious buildings near these local parks can enhance their performance and use. Florence, Italy. Authors' Personal Archive, 2022

Principle 3: Creating diversity and interest
Principle 5: Creating a city center based on local facilities and resources
Principle 7: Ensuring visual and functional continuity
Principle 8: Maintaining relative enclosure with human scale
Principle 9: Maintaining city centers' concentration and compactness
Principle 11: Security and safety of city centers.

6.3.2 Applying Architectural Solutions to Buildings in City Centers

Henry Faison believes that People do not come to the city center because of good architecture, but they avoid ugly and bad architecture. In the 1980s, the criteria for good and bad architecture in the city center changed (as noted in the previous chapter). This is eloquently expressed in the sentence, "The environment, not the architecture of its buildings, should be spectacular" (McBee et al. 1992). The character and vitality of the city's sidewalks are influenced by the buildings that form streets and public spaces in the city's core. In order to create a clear organizing structure and a high level of functional integration in the city center, these buildings must be connected to the street and to one another in a positive manner. Analysis of the existing architecture identifies commonalities in the placement and design of buildings, as well as how they contribute to the formation of good mental images for pedestrians. This understanding of the context is the starting point for building renovations and the construction and design of new structures. The following are significant roles architecture can play in the city center (Paumier 2004):

- Creating a continuous street front that is regarded as the first path of movement and provides the structure necessary for a unified city. The walls created by buildings clearly define the street space. Consistent setbacks and orientation toward the street strengthen the viewer's perception of the street as a delineated space. It is essential to use construction to fill vacant urban spaces, particularly on major sidewalks and arterial streets. Vacant lots, surface parking, and deep setbacks weaken the three-dimensional framework of streets and buildings by creating gaps in the street facades that open parking lots. In contrast, consistent front yard setbacks and infill development can preserve and strengthen the urban fabric and unify the city center's physical structure.

- Creating visual coherence through similarities in height, scale, massing, the general organization of facades, and the materials, colors, and shapes of roofs in buildings.
- Recurring themes in design help to create a recognizable identity and a sense of place for the city center, turning it into an attractive and suitable location for business activities. This does not mean that all buildings should be alike or that all the details and materials must be selected from a limited set, but that emphasis should be put on the common features that make the existing architecture special. An effective way to do this is to develop or restore a set of local or regional architectural concepts and their materials and prevent the undermining of the city center's special identity due to an architecture that is non-native or inconsistent with the context.
- Creating a sense of human scale, activity, and appeal at the street level to increase the desirability of the city center for pedestrians and to improve the integration of individual projects and the development of blocks. Multiple uses should extend from the street level to the edges of sidewalks using zoning guidelines and regulations, development codes, and other controls. Using large windows and entrances at street level, the facades of shorter buildings should expose the street to the activity within the building. Through the use of outdoor spaces, such as placing chairs in front of restaurants and cafes, for example, activities can be transferred to the street environment (Figs. 6.12, 6.13 and 6.14).

It is not necessary to control every detail of the design of a renovated building to achieve integrated development. However, fundamental decisions regarding setbacks, building heights, entrance façade lines, overall façade organization, materials, ground floor planning, and shop façade design should be coordinated.

Fig. 6.12 Cohesive and integrated structure of the façade of city center buildings enhances the sense of movement in the urban space and guides pedestrians between key spatial junctions. Milan; Italy. Authors' Personal Archive, 2019

Fig. 6.13 To create a sense of unity in the facades of the buildings in the city center, it is not required for the buildings components to be identical; similarity in the geometric form, materials, and location of building elements can accomplish this. Venice; Italy. Authors' Personal Archive, 2022

Fig. 6.14 Human scale of buildings and furniture, attractive uses, and a large number of windows in surrounding buildings have made this local plaza in the city center an inviting and pleasant place for pedestrians. Venice; Italy. Authors' Personal Archive, 2022

(A) **Building Facades**

Urban facades are the means of social communication in the city, as they convey information, values and emotions to people. Therefore, the appearance of buildings can have various effects on urban space users. The city center is the first place people wish to see upon entering a city. On the other hand, according to Bahreini (2020), the message conveyed by building facades is the decisive and lasting first impression of the city for every newcomer. The influence and dominance of modernism, on the one hand, and the intellectual chaos and turmoil caused by alienation from original values, on the other, have produced an equally confusing, imprudent, decrepit, and meaningless image in our cities (Bahreini 2020). Urban facades include not only the primary architectural elements of the buildings, but also components such as coolers, ducts, signs, and other components added to the building wall for various purposes. Therefore, in the study of urban facades, one must first distinguish between fixed and added elements and then visually and, if necessary, functionally evaluate each one. Additionally, the general form and style of facades, details such as doors, windows, and openings, projections and setbacks, and building extensions such as canopies, signs, wiring, etc. should be evaluated based on harmony and balance, proportions,

Fig. 6.15 Designing and organizing the facades of the ground floor of buildings in the city center due to the importance of the uses located in this layer, more observation and as a result better perception by pedestrians and influencing their experience should be prioritized. Bologna, Italy. Authors' Personal Archive, 2019

symbolic aspects, beauty and firmness, maintenance and restoration, order and legibility, simplicity and complexity, continuity and diversity, and composition. Due to the greater impact, it has on the pedestrian experience, continuity at the ground floor level is particularly crucial. Here, the use of similar building materials and a similar volume and shape of structures should be encouraged. Similarly, a reasonable distance between buildings, a positive relationship in location and the ratio of openings in the facades, and the organization and creation of highly defined ground-floor and upper-floor facades should be considered (Fig. 6.15).

In fact, the purpose is to coordinate the design of city center buildings and to turn the separate facades of each block into parts of an integrated macro-architectural unit (McBee et al. 1992). A shared understanding of community conditions and goals provides a better basis for establishing city center reconstruction strategies. The basic principles for the renovation of commercial buildings include the following (Paumier 2004):

- Strengthening the integrity and unifying the design of building facades.
- Creating storefronts that add to the appeal, dynamism, and comfort of the street environment.
- Emphasizing design, material, and color compatibility to show adjacent buildings as a unified whole.

Creating an integrated view for a block and organize various architectural styles and details in a particular pathway requires a good understanding of the façade design framework. Such a framework consists of two main components: the upper façade and the store front. In addition, the lateral and back facades can also be used in some cases to complement the two main façade types.

- **Upper Facades**

Upper facades consist of cornices and fascia that include the building's front façade and upper shops, as well as the windows that make the architecture of the upper floors

Fig. 6.16 Integrity of the skyline, continuity of the connection line between the building's ground floor and upper facade, the protrusion of the roofs, the imaginary line along the windows, and the balconies all contribute to the facade's harmonious structure, Milan, Italy. Authors' Personal Archive, 2022

appealing. In addition, they feature columns that extend to the ground floor, bringing visual consistency to the façade and providing a framework for the storefronts. The building's consistency, architectural quality, and personality are a result of the upper façade and integrated architecture. Consequently, the design method, materials, and upper façade conditions play a significant role in defining the architectural style and establishing the relationship between the buildings within a block (Fig. 6.16).

- **Cornice and Fascia**

Cornice and fascia create a strong roof line and give the building a complete and finished appearance. If these components are missing or replaced with something else, they should be rehabilitated and rebuilt to emphasize the content of the original building design. New cornices should be designed to fit the overall volume of the building.

- **Facade Materials**

All mechanical equipment and devices, electrical equipment, and panels must be removed from the surfaces of the main facades, which must be clean and flawless. When possible, surface construction materials such as metal panels, tiles, and plaster should be removed from the wall's main surface to reveal its surface in detail. Particular attention should be paid to the removal of facade materials from shops that extend to the columns and walls of the upper facades, as they can have a detrimental effect on the main architecture and weaken the equilibrium between the main components. If new materials and paints are to be utilized, they must be compatible with neighboring structures and complement the design of storefronts. Such modifications should be restricted to structures whose exterior surfaces cannot be altered or whose facades are not architecturally identifiable (Fig. 6.17).

Fig. 6.17 Beauty of the building facades in the city center is not an accident; rather, it is the consequence of high precision in the design and implementation of façade details, adherence to urban design guidelines and codes, and careful maintenance and restoration of the façade. Strasbourg, France. Authors' Personal Archive, 2022

- **Windows**

The main windows of the upper floors of buildings should be modified to create a sense of scale and increase the visual appeal of the upper facades. Recreating or defining the main windows above shops has a significant impact on the architectural integrity of many commercial buildings. The ratio of the modified windows and the rhythm of their arrangement pattern should, as far as possible, follow the design of the overall facade (Fig. 6.18).

- **Columns**

The columns that form shop fronts and create important visual anchors for upper facades play an important role in creating an integrated architectural framework that, in turn, forms a visual diversity across the street. Where columns are removed or downsized, the architectural definition of the ground-floor façade is weakened, and the architectural balance of the upper floors seems insufficient. In such a case, the width of the columns and the distances between them must be corrected to make the façade more consistent. In wide buildings, the use of columns that separate the shop

Fig. 6.18 It is crucial to find windows with a valuable pattern in a city's historic core because, in addition to providing a place's identity, they can be used to create urban design codes and guidelines that organize and integrate urban views; Bologna, Italy. Authors' Personal Archive, 2019

fronts is recommended to strengthen the balance. To emphasize the role of columns in defining the architectural character of the upper facade, their surface materials must be the same as those of the facade (Fig. 6.19).

- **Storefronts**

The upper facade columns and sign frieze that separate display windows and building entrances from the upper floor architecture characterize the storefronts on the ground floor. The lower part of the façade provides physical and visual access to commercial activity inside the shops and creates a space where the character and identity of the commercial uses can be best expressed. This part of the view is seen and experienced more directly by pedestrians.

Many shops in city centers feature elements often found in commercial structures. Repetition of such standard pieces throughout the street promotes visual cohesion and must be maintained (Winter & Company 2002). The storefronts that front the street are the focal point of the façade and can produce visual appeal and vitality, which in turn makes the street inviting. The storefronts are an intrinsic part of the entire block façade, making a solid horizontal connection with their recurrent display windows, synchronized symbol frieze, and building-connecting-colored canopies (Fig. 6.20).

- **Display Windows**

Modifications to the lower façade should emphasize the shops' interaction with the street and their transparency. The storefronts should be perceived by the visitor as an open space, generating a dynamic visual focus that contrasts with the dense upper façade volumes. A continuous band of display windows and entrances at street level invigorates the street environment and is a crucial part of the block façade. Display windows should never be obstructed or concealed, and altered windows must be returned to their original size. These windows should not have reflective or gray glass installed (Fig. 6.21).

- **Entrances**

Shops' entrances are their focal points. In older buildings, entrances with a slight setback can better define shop fronts, increase architectural variety for passersby, and

Fig. 6.19 Upper-floor columns and engaged columns establish significant visual anchors and while grabbing the viewer's attention accentuate the building's entry. Milan, Italy. Authors' Personal Archive, 2019

Fig. 6.20 Ground-floor storefronts are the most significant façade layer due to their visibility to pedestrians, the relevance of the ground-floor layer's functions, and the connection of the majority of the building's activities to the street. France; Strasbourg, France. Authors' Personal Archive, 2022

Fig. 6.21 Display windows are a significant component of the aesthetic attraction of city center streets, particularly for pedestrians, due to their unique and ever-changing artistic displays. Champs Elysees, Paris, France. Authors' Personal Archive, 2022

serve as a climate change mitigation factor. When entrances and display windows are aligned, awnings can be used to add variety. The entrance doors must be made of glass to maximize shop visibility. Their design style and material should correspond with the commercial design of the store. In residential areas, integrated doors should be avoided. Where upper-floor entrances are adjacent to the primary storefronts, they must be incorporated into the architectural framework so that they appear as secondary elements (Fig. 6.22).

- **Awnings**

Awnings are regarded as a simple, inexpensive, yet effective tool for enhancing commercial facades and creating a favorable impression for storefronts. They are a means of adding color, variety, and aesthetic appeal to the streetscape, as well

Fig. 6.22 Stone arches spanning the entire length of the street mark the entrance to the stores in the city's historic city core. Bern, Switzerland. Authors' Personal Archive, 2022

as increasing pedestrian comfort by shielding them from the sun and rain. Awnings should be utilized frequently to draw attention to the storefront and establish a strong horizontal element at the block level. Without additional columns on the sidewalk, they should be attached directly to the building. Hotels, theaters, and other significant structures can incorporate awnings into their architecture to reinforce entrances and shield pedestrians from the effects of climate change. Design guidelines should suggest a combination of awnings, signs, etc. that creates a pedestrian scale along the street (Fig. 6.23).

- **Signs**

Signs play a crucial role in determining the visibility and continuity of the city center. Consequently, the development of a simple and coordinated plan for the use and design of urban signs (used to indicate directions, road names, etc.) is also required to reduce visual clutter. Improving urban signs can yield rapid and inexpensive results that are both desirable and unimaginable. Additionally, seasonal or ceremonial banners, lighting, etc. should be utilized to add visual appeal to the streetscape.

Fig. 6.23 Red canopies around Siena's Plaza del Campo, in addition to enhancing the space's aesthetics and diversity, help to distinguish the cafes' break area in front of the stores. Siena, Italy. Authors' Personal Archive, 2019

Fig. 6.24 In some cities, hanging flags and colored cloth in the city center during festivities and historical and cultural events can foster a sense of civic pride and establish the space's character. Florence, Italy. Authors' Personal Archive, 2019

These measures can transform the city center into a frequented destination (Fig. 6.24) (McBee et al. 1992).

The appearance of individual structures and the streetscape as a whole is greatly affected by commercial signage. A sign typically serves two purposes: first, to attract attention, and second, to communicate information. If the building facade is well-designed, it can serve as an attention-getter in situations where it is highly visible, allowing the signs to concentrate on conveying information. All new signs should be designed with the wall's overall texture and the surrounding developed area's character in mind (Winter & Company 2002).

Signs are utilized to introduce a business, establish its image, and display its services. To be successful, they must be impressive and convey their message without providing unnecessary details or employing an excessive number of words. However, they should not be so abstract and succinct that they convey a message that is unclear and unintelligible. Signs can be designed and installed to complement building architecture, reduce visual clutter along the street, and function as an integral part of the block's facade (Fig. 6.25). The basic principles of designing an effective and appealing sign system are graphic simplicity and compatibility with the building's architecture. The characteristics of signs, including their size, location, material, color, calligraphy, and lighting, contribute to the creation of a positive identity for businesses and a unified image of the city's central business district. Consequently, it is necessary to create and implement urban design guidelines to integrate the dimensions, design, and placement of commercial signs, as well as a regular plan for the use and design of public signs and their display.

Size. Each sign should be proportional to the size of the storefront and the entire structure, without obscuring its primary architectural features. The ratio of one square foot (0.09 m^2) per foot (30.48 cm) of the width of the building facade is a good rule of thumb for determining the size of the signs on the building facade. The sign area on the building's facade cannot exceed 1.9 m^2 (20 ft^2) (Paumier 2004).

Position. Signs are typically placed on the lintel or sign frieze that separates the ground floor from the upper facades in traditional multi-story commercial buildings. In this location, the sign serves as a boundary between the two primary facade components and helps to define them. Wherever possible, signs for businesses that

Fig. 6.25 Shop signs in the city center should be very simple while effectively conveying the message and introducing uses or services, and they should not be overly impressive or attention-getting in terms of size, material, color, etc. In the case of historic buildings, this matter is approached with greater care. Siena, Italy. Authors' Personal Archive, 2019

share a front should be positioned at the same height to create a unified appearance (Figs. 6.26 and 6.27).

Sign Materials. The main point in choosing sign materials is their fit with the overall architectural character of the building. Many materials, including wood, metal, plastic, neon, and fabric, are suitable for making urban signs. The use of materials such as substandard plastic panels that induce low quality should be avoided.

Fig. 6.26 Proportions, proper color, and coordination of the signs set perpendicular to the pedestrians' line of sight aid in correctly communicating the information without disturbing the order of the space. Milan, Italy. Authors' Personal Archive, 2022

Fig. 6.27 Two distinct types of materials are used in the design and construction of the sign on this façade. The upper white sign is older, made of stone, and appropriate for the building's façade, while the lower sign is newer, made of metal, and distinct from the background material. However, the general design of the lower sign in terms of size, color, and location did not disturb the composition of the façade, and the negative effect of inconsistency between the sign and the building's materials was mitigated by the observance of proportions. Siena, Italy. Authors' Personal Archive, 2019

Color. Graphic design and colors can enhance the pedestrian experience of walking along the street and its commercial area. While bright colors are generally good for signs, the use of too many colors in a sign can be confusing.

Message. The message can be displayed in the text, image, volume, or a combination of these. Although the abstract method is used to convey messages, for public utilities and services such as telephone, post, firefighting, and law enforcement, the information should be provided explicitly and clearly (Bahraini 2003). The words on the signs should be limited to the name of the business activity and other information related to that activity. Simplicity is the key to readability, elegance, and beauty. Simple and large calligraphy and the use of well-known symbols are among the most effective techniques to this end.

Lighting. Signs that emit dazzling light or move around are often not suitable for use in the city center. Therefore, indirectly lit signs are preferred. Most storefronts can be lit by streetlights on sidewalks. The lighting of historic buildings and structures can contribute to the quality of city center spaces. Private sector builders and their architects must use special façade lighting to enhance the public realm (Fig. 6.28).

- **Side and Rear Elevations**

Such facades, when accessible from the street level, can have a significant impact on the mental image and visual character of the city center. Though many facades do not have high design quality and some building fronts are left incomplete, they can be improved to provide a more attractive and organized appearance.

Fig. 6.28 Use of fixed lighting with appropriate lighting levels, while emphasizing the desired advertising message and maintaining the quiet and calm spirit of the street, has helped to increase the sense of mental security at night. Lugano, Switzerland. Authors' Personal Archive, 2022

Fig. 6.29 Rotation of the facade and the build-to-line[1] in accordance with the corner building has resulted in the continuation of the street view of the windows, store lines, and upper floors. Rome, Italy. Authors' Personal Archive, 2019

- **Corner Buildings**

As corner buildings show the overall quality of the block, their storefronts should be chamfered at the corner so that the continuous streetscape is not disturbed. In such buildings, the overall design of the side facades mimics the architecture of the main building facade (the building front) and indicates the progression from one area of the city center to another. Corner facades should adhere to the same design guidelines as storefronts and upper facades that are being maintained or altered (Fig. 6.29).

- **Incomplete Side Facades**

If unfinished facades are visible from the street, their quality should be enhanced by removing or concealing exposed mechanical equipment and increasing the use of materials, colors, or details on the building's front walls. Where windows cannot be

[1] Build-To-Line is the line parallel to the street along which the primary mass of the front façade should be set. It is measured as a perpendicular distance from the street line to the nearest point of the building façade.

Fig. 6.30 The mural of a national, religious figure praised by the civil society is done in a part of the city center in a facade without openings, in harmony with the context. Mural painting of San Gennaro, the patron saint of Naples, greets pedestrians between these narrow city streets. Naples, Italy. Authors' Personal Archive, 2022

created, graphic design can be used to make wall surfaces more appealing, and it is typically more effective when designed on a neutral-colored background (Fig. 6.30).

Rear Facades

When parking lots are located behind buildings in the city center, they can be used as secondary entrances; consequently, they must be designed in a way that is both aesthetically pleasing and creates an identity commensurate with the building facades. Each wall surface must be meticulously repaired. Service areas and warehouses must be meticulously organized and maintained. Closed windows must be reopened, and entrance doors, billboards, and lights must enhance the facade's aesthetic appeal. Awnings, window graphics, and landscape design can also be used to enhance the rear facades of a building (Paumier 2004).

(B) New Developments

New developments should be designed in such a way to complement the existing architecture and enhance its features.

- **Infill Developments**

These developments can strengthen and improve the urban context by eliminating gaps created by vacant lots and open parking lots. After protecting the good architecture that is already there, these empty spaces should be used first for building. The pedestrian spines, primary connectors, special public transportation streets, and main arterial roads should be key target locations for such developments.

Infill buildings should be designed with special sensitivity in a way that strengthens the positive features of the existing architecture. Contextual design is not just to apply the architectural style of a period to existing buildings, but to bring back the main features of the existing design with an acceptable contemporary expression. Under the conditions in which traditional bazaars were developed before the twentieth century, the scale, form, orientation, and main materials of buildings were almost fixed, with variety and contrast being created mostly through differences in details and decorations on a small scale. As a result, general consistency and character were easily achievable. However, technological advances such as elevators, air conditioning systems, metal building structures, and modern architecture offered a wider range of choices in terms of form, scale, materials, and character of buildings. The result was a greater potential for contrast. In an effort to create a distinctive identity for a particular project, newer urban architecture has turned a blind eye to consistency in favor of contrast. This extremist approach has often led to disorder and incompatibility in the urban environment.

A combination of old and new buildings can add variety, appeal, and depth to the visual character of the city center (Fig. 6.31). Similarly, sharp contrasts in the design of buildings can be exciting and prominent if they create meaningful signs and focal points. But to be effective, buildings with contrast must be in a relatively uniform and calm architectural context. An important point is that the threshold of tolerance for contrasts declines as the size of the city decreases. In smaller cities, sharp contrasts must be carefully created to avoid disrupting the existing architectural order. In order for infill buildings to remain consistent with existing constructions, they should emulate key aspects of existing development rather than imitate them (McBee et al. 1992).

- **Front Setbacks**

New construction should adhere to the building setbacks of existing structures so that building edges are connected and coordinated, urban development patterns in the urban core are strengthened, and pedestrian orientation is enhanced. In the urban core, setbacks are typically adapted to public movement routes and allow ground-floor active uses to open directly to pedestrian areas (Paumier et al. 1988) (Fig. 6.32).

- **Space Between Buildings**

Setbacks of buildings with side yards should repeat the rhythm of existing building spacing. In the developed core of urban centers, building setbacks with side yards are usually eliminated elsewhere, except where sidewalks have been created between blocks. To be effective in creating a pattern of intensive urban development, the front

Fig. 6.31 Continuation of the main lines of the facade and the repetition of the key design principles of the older building's facade in the new buildings constructed using the infill development strategy have resulted in the creation of a cohesive and character-driven environment. Milan, Italy. Authors' Personal Archive, 2019

Fig. 6.32 Setbacks facing the main street can take various forms. In the meantime, the construction of arcades and colonnades on the ground floor of buildings and along the main pedestrian movement axes is a common example of setbacks in European cities. This type of setback can have similar functions to the common type of setback (retreating the building line from the continuous build-to-line) in order to create an inviting space at the entrance of a significant building or define a space for a brief rest. Bologna, Italy. Authors' Personal Archive, 2019

and side setbacks of buildings should be reduced as they approach the city center. Also, front setbacks must be deep enough to create wide and landscaped sidewalks to humanize the pedestrian environment.

- **Height and Volume of Buildings**

Although leveling the height of buildings can be an important factor in creating a sense of visual continuity, it cannot be definite or absolute because of a degree of visual error, which is natural in human vision. However, precise transitions between buildings of different heights can help reduce discontinuity and integration. The height and volume of new buildings should be commensurate with existing developments and create calculated height transitions between low-rise developments and new high-rise buildings. The volume of buildings must also be broken down into smaller parts using various setbacks and roof heights so that they can adapt to the scale and volume of existing buildings.

- **Entrances of Buildings**

The entrances and main facades of buildings should be oriented toward streets designed as pedestrian spines, primary connectors, or major arterial roads. The spacing and articulation of entrances should reflect what is seen in existing buildings (Paumier 2004) (Fig. 6.33).

- **Façade Organization**

As seen in the traditional architecture of commercial buildings, the front facades of infill developments should be organized into two major components: the ground-floor facade of shops and the upper-floor architecture, separated by strong horizontal elements. Especially at pedestrian crossings, where a sense of human scale and the presence of amenities are crucial, the ground-floor facade of shops should include

Fig. 6.33 Opening the main entrances of buildings to the main axis of pedestrian movement has increased the significance of this plaza in the heart of the city. In addition, the continuity and integration of the continuous colonnades and arcades along the parallel commercial line of the plaza, while increasing the spatial order of the entire facade, have resulted in the creation of a character that is both independent and integrated. Milan, Italy. Authors' Personal Archive, 2019

Fig. 6.34 Use of strong horizontal lines between the ground floor and upper floor has divided the wall's volume into distinct sections and increased the variety of discernible forms on the facade. In addition, a dynamic pedestrian edge has been created for the establishment of commercial uses in this layer through the use of large and transparent openings on the ground floor. Geneva, Switzerland. Authors' Personal Archive, 2022

large windows so that they can share the activities inside with the street. The street-facing facades of multi-story parking garages should never be occupied by parking spaces (Fig. 6.34).

- **Building Colors and Materials**

Each color has its own meaning and concept, with proportions that are calculated and reasonable, indicating order and discipline. No color is employed unless it communicates a message and accomplishes a purpose. This language also has regional and indigenous dialects. Thus, the meanings and concepts of colors can vary and even be contradictory in different cities, resulting in a unique identity for urban areas or cities (Shakibamanesh 2015). The variance in hues of city center structures is largely attributable to the climate and the building's construction materials. Therefore, each area's color must be adapted to these conditions. Red bricks, for instance, are very popular in Boston due to their prevalence in existing structures. In order for building facades and architectural decorations (details of windows, doors, cornices, etc.) to be optimally visible, it is best not to use the same paint colors for both (Fig. 6.35).

A possible exception would be the use of monochrome coloring where historically color has been used. The use of improper colors and materials in the construction of a building or its alterations can result in a great deal of context inconsistency. When similarity to old buildings is not required, new materials in the city center should be coordinated with existing structures in terms of context, scale, and color (McBee et al. 1992).

Fig. 6.35 Use of colors on the street facades within the frame of enacted plans is advantageous in a number of ways: A-establishing a sense of affiliation and belonging in residents; B-enhancing the vitality of pedestrian axes by employing the psychology of colors; and C-providing a distinct identity to certain urban axes based on their functional or cultural characteristics. Istanbul, Turkey (Above); Venice, Italy (Bottom). Authors' Personal Archive, 2017 & 2022

- **New Major Structures**

The main challenge in designing large buildings in the city center is integrating large-scale and high-rise structures with existing small-scale structures. In a street or an area with attractive and coordinated architectural features, there must be an appropriate level of compatibility. The dominant scale and setbacks of existing buildings should follow a framework into which the new architecture fits well. However, when the existing architecture is at a moderate level, the first new project can create a basic framework for other buildings to create a new context. Certainly, where existing design standards are not strong enough, duplication of design components will not be desirable, and the new building should be used to develop new standards (Paumier et al. 1988).

In new major structures, buildings should be subdivided into smaller parts to complement the existing urban fabric and create transition points for height and scale. Moreover, the use of incompatible forms and volumes should be avoided. Very large multi-block buildings can harm the basic organizational structure and the effective flow of pedestrians toward the city center.

- **Setting Appropriate Height Restrictions**

In determining maximum height standards for new constructions, the city center development goals should be considered first. Such considerations include:

- The purpose of a particular area (is it the rehabilitation of existing structures or redevelopment?) and the appropriate tools that can be developed to protect specific structures or areas;
- The priorities that should be considered for the horizontal distribution of market potentials along the street to create a quality environment;
- The economic aspects of development that are influenced by factors such as the difference between existing and new densities, land cost and its impact on development possibility, and the market's ability to absorb spaces that can be assigned at the right time;
- The ability of developments to meet the increasing traffic volume and parking demands without compromising the quality of the street environment;
- The use of design guidelines for high-rise buildings to reduce visual effects, shading, and wind effects in new developments.

If these goals are not taken into account in the final decision to determine the maximum building height, the decisions in this regard will become more personal and political. Such decisions will also continue to be influenced by the tendency to strike a balance between economic development and land-use planning goals. These goals include increasing employment and income taxes and creating a large volume of activities in the dense central core in order to generate a human scale and a desirable identity for society. Therefore, it should not be forgotten that the way the street surface is designed is as important as the building height in creating the mental image of the city center.

- **The Location of High-Rise Buildings**

When the city center's architecture creates a desirable identity and human scale in the pedestrian spine and primary connectors, it may be preferable to add denser high-rise buildings to those in the middle of the block or near the traditional core.

New high-density buildings should be constructed within a convenient walking distance (less than 9 min) of conventional commercial units. In many instances, these buildings play a greater role in bolstering the economic vitality of city centers than in establishing competing centers (McBee et al. 1992). In both instances, the new high-density buildings must be placed at a distance from the traditional shopping center in order to facilitate the movement of pedestrians, increase the vitality of the city center, and prevent the emergence of competing city centers. Transition points in the city center's heights and volumes should be meticulously designed. For example, height-to-setback regulations can reduce the visibility of tall buildings from historic street fronts (Fig. 6.36).

- **Controlling the Effects and Consequences of High-Rise Buildings**

The psychological effects of building height can be mitigated by constructing a strong horizontal element that acts as a "roof" for the street. Lintels and setbacks that define the effective height of the facade from street level must be at the same height as cornice lines or lintels on new buildings.

Fig. 6.36 Creating a modern central business district by locating and constructing high-rise commercial buildings at the right distance from the traditional city center. Milan, Italy. Authors' Personal Archive, 2019

Forms that create a three-dimensional reference scale and articulation on the ground are effective for creating a sense of human scale at the street level for new developments, including high-rise buildings. Because the base of the building and its line of intersection with the ground must serve to humanize the street environment, reflective and extensive glass should be avoided on the ground-floor facades of buildings. To improve the visual appeal of high buildings on the skyline, sculpting their tops can be included in design guidelines. In addition, guidelines for increasing sunlight exposure in passages and public places should also be provided, and measures should be taken to reduce the effects of wind tunnels or downdraft effects.

- **Megastructures**

The term "megastructures" is used to refer to large-scale building developments, whether they are small, tall, or a combination of buildings with different heights occupying one or more blocks (such as a new shopping mall or major mixed-use developments). Integration of such large-scale developments with their surrounding environment should be given special consideration. When these projects connect multiple blocks, they can disrupt the primary pathway network, which provides the city center with a coordinated development pattern, and sever street-level connections between key locations. The design of megastructures should preserve the connection between pedestrians and visual corridors and expose their interior activities to the street. Architectures with buildings that turn their backs on the city, create empty continuous walls on the street, or primarily focus on indoor activities, should be avoided. The most important recommendations to reduce these negative effects include:

- Combining existing buildings and facades in large complexes;
- Breaking continuous and long facades into smaller sections through the use of windows, architectural details, various setbacks, and rooflines, and creating a human scale on the traditional streets of the city center;
- Articulating building volumes into a set of smaller components to reduce the mental image of a hard volume;
- Creating sequence as seen from public places and sidewalks connected to the street network;
- Orientating major views and entrances toward streets that act as the main corridors for pedestrians;
- The use of ground-floor facades of shops and commercial activities to adapt building functions to the street; and
- Designing transitions between the height and volume of large projects and existing or planned structures in their neighborhood.

The design guidelines developed for such new structures allow developers and builders to meet the goals of urban planners to create a coherent pattern of development and a more aesthetically pleasing environment. Compliance with these design guidelines reduces disagreements over proposals, improves the overall quality of the design, and expands architectural rational action freedom. The purpose of these guidelines is not to discourage investment and innovation, but rather to ensure that the development of large-scale projects does not harm the character and future prospects of the city center.

(C) Renovation and Preservation Solutions in City Centers

- **Renovation and Reuse of Buildings**

Human scale, quality materials, and architectural details of old buildings contribute to the attractiveness and identity of the environment. Wherever possible, examples of the traditional architecture of commercial and residential buildings in the city center should be preserved, renovated, and adapted to suitable locations to accommodate new uses. By being repurposed as offices, residential units, and commercial spaces, old and abandoned industrial buildings are given new life (Fig. 6.37).

Renovation and adaptive use of historically significant and aesthetically pleasing buildings contribute to a favorable environment for reinvestment and restoration by enhancing market access. Rehabilitating dilapidated, abandoned buildings and poorly modernized storefronts, as well as modifying old paintings, can assist the center in preventing the creation of an impression of abandonment and stagnation. While the

Fig. 6.37 Adaptive reuse of a valuable old building by preserving the design values in architectural details (such as the store's ceiling) in the form of an attractive and dynamic contemporary use can result in the return of citizens and the improvement of economic conditions essential to preservation. Lego store, Florence, Italy. Authors' Personal Archive, 2022

presence of abandoned buildings in the city center indicates an economic decline, the renovation of highly visible old buildings represents the reconstruction movement and the impulses of revival and creates an image of evolution and renewal (Paumier 2004).

- **Preservation**

Historical preservation is a means of contributing to the creation of a spectacular environment and of utilizing the character and identity of historic architecture for economic purposes. People appreciate working and shopping in environments that provide exceptional personal values, qualities, ideas, and services (Fig. 6.38). Traditional (historic) buildings in a city center can make them distinctive. Therefore, historical preservation necessitates modifications to the management of the built environment due to the rapid expansion of urban communities and the need for their architectural styles to evolve. When new, well-designed buildings integrate with existing architecture, a rich pattern of visual cues emerges that reflects changing cultural and social values (Smith et al. 1996).

Preservation does not imply opposition to change (see Chap. 4), and does not recommend the rebuilding of historic buildings, exclusive use of historical themes in new buildings, or the preservation of all historic buildings. Rather, it recognizes and extracts good design patterns from past locations that are evidence of extraordinary past events. In addition, it preserves the unique features and characteristics of built

Fig. 6.38 To return the investment in the historical contexts of the city center and help cover the cost of maintaining and protecting the architectural heritage, it is crucial to utilize the potential and capacity of valuable spaces and buildings with historical identity and adapt their use to meet the needs of contemporary citizens. Rome, Italy. Authors' Personal Archive, 2019

environments as well as good designs that are consistent with the past (during the construction of new buildings or the delineation of existing structures) (McBee et al. 1992). Coordination between new buildings and changes to the historical context is crucial, but it is undesirable to merely imitate older structures. In fact, buildings with a distinct style from older structures in the city center are preferred, so long as they reinforce the overall design of the old development patterns and do not harm the architectural identity of their context (Winter & Company 2002).

The preservation and maintenance of old structures can play an important role in protecting the distinctive identity of a city and fostering a sense of history and human scale in cities facing increasing pressures for city center development. To date, numerous strategies and controlling factors have been implemented to safeguard buildings that contribute to the historical character and architectural identity of urban cores. Where redevelopment pressures are intense, special measures are required to prevent the unwarranted depletion of such resources. The following guidelines should be considered by those responsible for the development of city centers (Paumier 2004):

- Existing zoning rules and regulations should be carefully reviewed to identify areas where pressures for increasing density and new developments threaten historic places. A low-rise historic building in a place where high-rise buildings are allowed can be seen as valuable social capital.
- Where priority is given to building maintenance and reuse, the introduction of special landmarks and historic sites for preservation can alleviate the pressures of new construction by forcing revision and destruction permits in these areas. Using the revision process to make changes to the architecture of a building, the definition of such valuable buildings and areas can be a good way to coordinate those changes.
- For preservation and repair, you need incentives and financial help, like tax-related facilities and low-interest loans.

- In some cases, creating public shopping opportunities is essential for the preservation of historic sites. Because of this, it can be helpful for both users and owners to take advantage of long-term and short-term leases that make reusing and fixing up these buildings economically worthwhile.
- Transferring building permits to other parts of the city can make it easier to take care of and keep historic buildings in areas where zoning plans call for a lot of new construction.

A building's preservation and renovation may not always be appropriate. On the basis of architectural research that specifies historically significant and influential structures, concentrated preservation efforts can be pursued most effectively. Typically, these structures are situated in the heart of the city's traditional market and commercial center, particularly along the main shopping streets in city centers and in residential neighborhoods. Inactive and non-contributing structures in the city's central core can provide suitable locations for the expansion and construction of new structures. When carried out concurrently, the advancement of new developments and the preservation of old influential structures can generate investment opportunities and strengthen the city center's economic standing in the region.

Moreover, it is essential to determine whether the renovation of certain old buildings is economically viable. This measure necessitates a deliberate evaluation of the historical characteristics and economic justification of these buildings. Destruction may be the best option when the rehabilitation of a derelict building discourages investment in the surrounding area. The public sector should take serious measures to prevent the destruction of buildings if a reasonable reuse proposal is not received within a reasonable time period.

6.4 Solutions for the Traffic System in City Centers

6.4.1 Which Principles of Designing Successful City Centers Are Met by These Strategies?

Application of solutions for the traffic system in city centers can help achieve the following principles out of the 15 principles outlined in the previous chapter:

Principle 1: Development based on an organizing pattern
Principle 7: Ensuring visual and functional continuity
Principle 8: Maintaining relative enclosure with human scale
Principle 9: Maintaining city centers' concentration and compactness
Principle 10: Increasing accessibility and pedestrian and vehicle access to the city center
Principle 11: Security and safety of city centers
Principle 12: Comfort and well-being

Principle 15: Balance between the static and dynamic aspects of pedestrian movement.

6.4.2 Applying Traffic System Solutions

(A) **Street and Block Patterns**

The street system is considered the backbone of the urban design framework of the city center. The extent to which such a system can create a predictable, unifying pattern of same-size blocks and regular spatial intersections determines the power of organizing structures for city center development. Block size has a significant effect on the scale of buildings. Blocks with a relatively small size found in older parts of city centers (sometimes with a width of about 60 to 90 m) also limit the width of the buildings. Small blocks also provide comfort for pedestrians by providing them with easy access and a view of other blocks and subdistricts.

In The Life and Death of American Cities, Jane Jacobs devotes a whole chapter to explaining the benefits of smaller blocks. Jacobs explains that small blocks make it possible to have a fine network of streets and provide many opportunities to choose different paths when passing through urban areas. With a greater variety of access routes, a pedestrian can use different routes over time and see more places, allowing him/her to have a sense of belonging to a larger part of an area. The presence of large-scale blocks and the lack of possible paths in a central area increase the number of areas on the street that a citizen rarely passes through. As a result, the feeling that these areas are important in his/her life fades away, and the chance for social sense and collective interaction is lost (Jacobs 2005).

Although small blocks may not be the best model for maximizing traffic flow or minimizing development costs, they can certainly be more effective than super blocks (90 to 180 m long) and giant structures in shaping people's experiences of the city. When urban decision-makers want to intervene in the existing street pattern by prescribing a block size larger than the existing ones, they should also pay attention to the fact that this decision undermines the essential element in defining the structure, scale, and functional coherence of the city center (Paumier 2004).

(B) **Hierarchy of Streets**

The grid street system, which is associated with the conventional pattern of urban development, is still the most prevalent layout in city centers. Although this system is easily memorized, it can become tedious and monotonous in situations where all streets are identical. To achieve a sense of hierarchy, the width and design of the sidewalks must be varied, and the street's facade and landscape must be altered.

Various roles of streets in the grid pattern can improve the visual structure of the city center, provided that the functions of streets are reflected in a similar hierarchy of the streetscape. Hierarchy coupled with the correct definition of major arterials, transit ways, collector roads, and local access streets can reduce possible collisions between local and transit traffic and increase the efficiency of the movement system.

This hierarchy can reduce the traffic volume and the speed of cars in the development of pedestrian-centered streets (Paumier 2004).

(C) Parking in the City Center

Planning parking spaces is crucial to making the city center a desirable place for people. Unless parking lots are well-designed, the city center cannot be a pedestrian center and host a variety of uses or a continuous flow of street-level activities. Open parking lots can create large gaps in the structure of development and separate important and essential uses.

It is essential to provide sufficient parking, but it is equally important to reduce the amount of land required for this purpose, particularly in the city center, where commercial centers currently exist or can be developed in the future. Creating temporary and short-term parking spaces for customers takes precedence over creating permanent parking spaces for employees. To minimize the impact of car parks on the visual quality of street views, car park facades should be restricted to major streets and significant sidewalks. If at all possible, parking lots should be situated on the lower levels of buildings or within interior-block structures behind active uses.

- **Limiting the Use of Long-Term Parking**

Several cities have found that they can support the expansion of office uses and, at the same time, reduce the need for long-term parking for staff in the highly developed central core by implementing parking policies and proper public transportation management. For example, in larger cities, the cost of long-term parking in the city center can be increased, and the provision of parking for daily intracity trips may be limited so that people resort to public transportation. However, these major changes in parking policies must be accompanied by measures to improve public transportation within the city. Otherwise, the city center's ability to attract and retain office and commercial uses will reduce as people are forced to park their vehicles far away from the city center.

- **Encouraging the Creation of Multi-story Car Parks**

The size of the area to be allocated to parking in the city center and its impact on the city center and surrounding areas can be reduced by creating multi-story car parks. Land value and construction density should be high enough to cover the cost of building multi-story car parks, which is far more than the cost of creating open parking lots. Many smaller cities may afford only one multi-story car park with a strategic location.

- **Size and Location of Parking Lots**

If the identity of the city center as a multi-use destination is made more apparent, the perceived deficiencies in the size and location of city center parking can be mitigated. A compact central core with well-defined, strong connections between activities and a pedestrian orientation is regarded as a "one-step" activity center where people can perform a variety of activities on foot after parking their cars once. The use of comprehensible signposts, maps, and other means of illustrating important routes

Fig. 6.39 Using the space under the stairs to construct the parking needed by residents and users is an acceptable idea, but the location of these parking places on the ground floor's valuable view has ruined its aesthetic appeal. However, this negative effect has been partially mitigated by the retreat of the upper-level buildings. Bern, Switzerland. Authors' Personal Archive, 2022

and locations can alleviate much of the confusion experienced by first-time visitors to the city center.

As the commercial backbone of the city center, pedestrian pathways for commuting between parking lots and the main activity centers of the central area should be made attractive and appropriate. Private sector builders can accomplish this by incorporating active uses into interior-block access pathways. Behind the shopping malls in the primary shopping district of the city center, the facades and entrances of open parking lots should receive special attention.

The size of parking lots should be proportional to the volume of traffic on adjacent streets. If at all possible, these parking lots should not be located on major pathways and streets with high pedestrian traffic, so as not to detract from the significant role that these pathways play in defining the appearance of the city center and the flow of pedestrians (Fig. 6.39).

- **Considerations for Parking Design**

City center parking lots should be designed in such a way as to enhance the quality of the visual environment and minimize their negative impact on commuting among commercial uses as well as between the city center's core and other activity centers.

On-Street Parking

Short-term parking on the street can assist in revitalizing the city center. The classification includes both angled and parallel parking. Some individuals prefer parallel parking in urban environments because they believe it saves more space and makes parking easier. However, angled parking lots are regarded as a detriment of marginal development. In an urban setting, parallel parking lots are preferable because they enhance the appearance of the street and reduce the traffic risks associated with angled parking lots. Although on-street parking is generally desirable for short-term

Fig. 6.40 Order and good arrangement of on-street parking on a minor street of the city center, Bologna, Italy. Authors' Personal Archive, 2019

use, it is sometimes necessary to eliminate it in urban cores in order to accommodate pedestrian amenities and improve street views. In some instances, reducing the number of carriageway lanes enables the maintenance of on-street parking, thereby improving the pedestrian environment (Figs. 6.40 and 6.41).

Surface Lots

These parking lots reduce the quality of the developing edges by interrupting the activities that make the street lively and appealing. Extensive sidewalks and parked cars create a visually harsh environment, which detracts from the mental image of the city center. The expansion of surface lots can weaken the stability of these areas rapidly and reduce the effects of land use mixing in boosting vitality and strengthening the city center as a regional activity hub.

Fig. 6.41 Considering space as on-street parking in close proximity to the wall, or on the sidewalk is not recommended. This activity can diminish people's perception and mental image of the ground floor, destroying the street's dynamics. Zurich, Switzerland (Right image), Bern, Switzerland (left image). Authors' Personal Archive, 2022

From the street, therefore, open parking spaces should be located in such a way as to reduce their visibility from main streets and crosswalks to a minimum. Landscaping the environment with plantings and shady trees can soften and even keep these harsh environments out of sight. A buffer wall or hedge with a height of 1.2 to 1.5 m can also be useful to reduce the visibility of parking lots. For security considerations, surface lots should be considered with a height difference of about 1.2 to 2.4 m lower than the surrounding ground level so that the area inside the lots is visible from the street level. The use of landscaped berms is not a suitable technique for separating the surfaces and creating buffers in the center of the city because they require a lot of space and induce a suburban image in the viewer's mind (English Partnership 2006).

Surface lot design standards must meet the need for plantings in interior landscaping. These planted areas define green islands by curbs and provide space for planting shady trees. These standards should also specify the amount of lighting and the types of lights compatible with those used in the city center and should be adjusted for the dimensions and location of the signboards and the body containing them.

Parking Structures

Parking structures require less land and can be covered more effectively to reduce their visibility. But they can also reduce the quality of the city center if they are located on the front line of the main commercial street in the city center. When these parking structures are on streets that contain primary pedestrian connectors, commercial spaces should occupy at least 50% of the space in front of them on the ground floor to minimize gaps in the activity and dynamism of pedestrians. In other streets where the ground floor of parking structures is occupied by parked cars, a landscaped setback should be used to soften their visual impact on the street and sidewalk (English Partnership 2006). Some cities have managed to build interior-block parking structures in the city center, providing enough space for multi-story development on the street front. Such interior-block parking structures should be encouraged because they promote the deployment of active uses along pedestrian-centered streets (Fig. 6.42).

Scale is an important factor in determining how a parking structure affects the urban environment. The length and width of parking structures on each street should not exceed 76 to 91 m and 55 m, respectively. Three-bay parking configurations are more desirable than other configurations (Paumier 2004). The architecture of parking structures should utilize the same materials as the adjacent building facades. Consideration should also be given to the use of neutral colors and architectural forms that complement neighboring structures. The facade facing the street of the parking lots should be designed in such a way that the fenestration pattern of the adjacent buildings is repeated in it, and in this way, this building is integrated into the architectural context of the central core. The design of a parking structure should under no circumstances feature unusual shapes, materials, or colors.

Fig. 6.42 Proper illumination, the presence of active uses near the entrance, and the minimum width of the lot keep the psychological security of this city center parking structure acceptable. Florence; Italy. Authors' Personal Archive, 2022

Concerns about security, which are perhaps more in the minds of individuals than in the realities of crime rates, can make people reluctant to use these car parks. Open or glass-walled steps, elevators with glass enclosures, and even more lighting in the building can increase peace of mind. In sensitive areas (for example, near residential areas), adjacent neighborhood areas should be protected from the light of parking lights. Underground car parks have the least visual impact on the city center and pedestrian areas, but they are considered the most expensive option (at about twice the cost of surface car parks). If underground car parks can cover the costs of their construction, it will be a worthwhile solution.

References

Bahreini SH (2012) Analysis of urban spaces: In relation to the behavioral patterns of users and criteria for design, 9th edn. Tehran University Press, Tehran, Iran

Bahreini SH (2020) Urban design process, 11th edn. Tehran University Press, Tehran, Iran

English Partnerships (2006) Car parking: what works where toolkit. Retrieved from: http://www.englishpartnerships.co.uk/carparking.htm/

Jacobs K (2005) The Manchurian Main Street: are shopping districts inspired by new urbanism a form of cultural brainwashing? Metropolis 24(10):110

LMN Architects (2005b) Mill Creek Town Center Design Guidelines. Mill Creek, Washington, DC

McBee S, Urban Land Institute, J. Basile, Ralph, Urban Land Institute Urban Development/Mixed-Use Council, C. Petersen, David, L. Phillips, Patrick (1992) Downtown development handbook. ULI—The Urban Land Institute, Washington, DC

Paumier C (2004) Creating a vibrant city center: urban design and regeneration principles. Urban Land Institute

Paumier C et al (1988) Designing the successful downtown. Urban Land Institute

Shakibamanesh A (2008) Promoting city centers quality by using special urban design guidelines (master's thesis). Iran University of Science and Technology, Tehran, Iran

Shakibamanesh A (2015) Fundamentals and principles of color design in architecture and urban design, 5th edn. Helleh Press, Tehran, Iran

Smith K, Joncas K, Parrish B (1996) Revitalizing downtown. National Main Street Center, National Trust for Historic Preservation

Winter & Company, Hoshide Williams Architects, Nakano Associates, LLC (2002) City Center Design Standards. City of Bellingham, Bellingham, Washington

Chapter 7
Urban Design Guidelines

7.1 Introduction

The physical form of a city cannot be determined by principles, policies, and strategies, nor by the urban development plan per se. Acknowledging environmental qualities and their promotion, as the ideal of the knowledge and art of urban design, cannot be confined within the framework of these documents and, given its complex nature, requires a different tool to implement its various aspects with more transparency. In order to control the quality of the urban environment, urban design guidelines play a crucial role in elucidating urban design documents. They offer guidance to urban designers, architects, and urban planners in order to facilitate the emergence of a development plan in accordance with the design principles and related policies and strategies. As part of the development document, these guidelines manifest development policies as key solutions and govern the qualitative aspects of the plan.

Urban design guidelines for the implementation of the fifteen principles (described in Chap. 5) and the solutions related to these principles (discussed in detail in Chap. 6) are addressed in this chapter under three broad categories: landscaping, building architecture, and access systems. In discussing the specifics of each category, the category's intended goals are stated at the outset, followed by a graphic depiction of the applicable guidelines.

A. Shakibamanesh and M. Ghorbanian, *Designing Responsive City Centers*,
Design Science and Innovation, https://doi.org/10.1007/978-981-97-8911-5_7

7.2 Urban Design Guidelines for Landscaping in City Centers

7.2.1 Street Landscape in General

Objective: To use landscaping measures to enhance visual and functional continuity, establish a desirable identity in the city center, and create an environment with a human scale.

Design guidelines:

(1) In urban design, the streetscape should create a unified image, a defined visual structure for the city center, and an inviting and convenient environment for pedestrians.
(2) The streetscape should create an attractive background for commercial activities and a proper environment for other activities in the city center through a visually desirable appearance (which does not reduce the attractiveness and visibility of the street front).

(3) The natural conditions of the land, including elevations, views, and other natural potentials, should be used to create a distinctive visual identity in city centers.

(4) All elements used in landscaping should be designed as a unit and combined, and their design and placement should avoid ignoring their effects on the overall character of the public territory in city centers.

(5) Intermittent use of the required street furniture, such as lights, pavements, benches, trees, plants, flower and newspaper kiosks, public telephone booths, etc., can be used to create a visual overlay and strengthen the visual and functional continuity of the organizing structure of the city center.

(6) If the city center contains specialized sub-districts, multi-purpose centers are created. Some representative landscape design elements in the city center (e.g., flooring, streetlights, trees, etc.) can be extended to sub-districts and at the same time retain a special character for their districts.

7.2.2 Pavement

(7) Continuity and consistency of the pavement plan should be maintained in every visual sequence.
(8) The pavement of new street areas should either complete or reflect the characteristics of previous pavements in the main street.

(9) The pattern, color, and texture of the paving should be in harmony with the surrounding buildings and even streets.

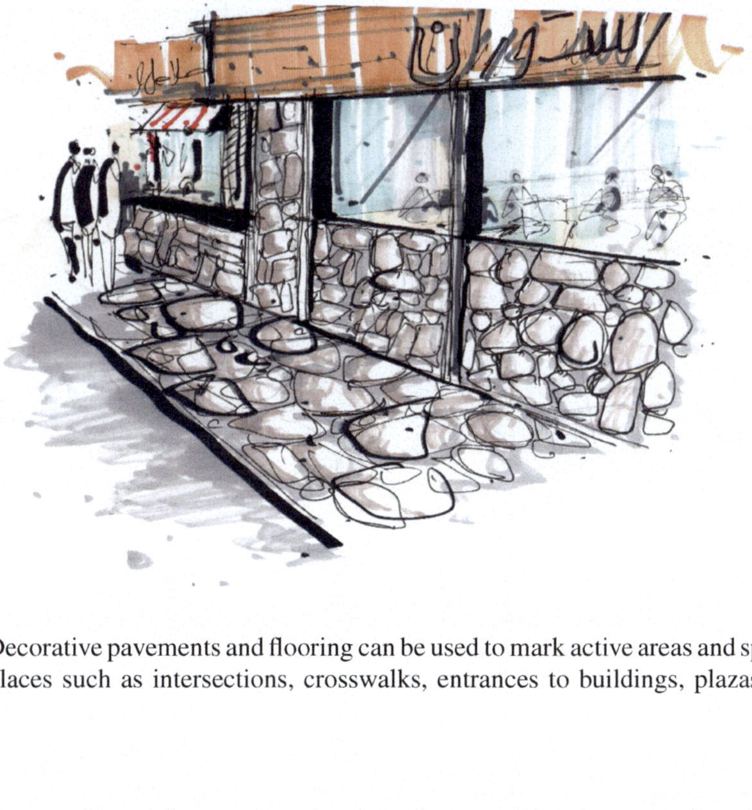

(10) Decorative pavements and flooring can be used to mark active areas and special places such as intersections, crosswalks, entrances to buildings, plazas, etc.

Moreover, such pavements and flooring can help define places for stopping, sitting, and other activities that can be performed in the space.

(11) Pedestrian crossings should be defined and qualitatively improved using landscaping measures such as the use of special pavements.

(12) Excessive use of pavements discourages pedestrians and reduces their activities. It also weakens the distinction of the edge of important activities on city center streets.

(13) Many porous pavements allow water to penetrate directly into the ground and reach the permeable layers of the soil. This method, referred to as "low impact development" (LID), is used to prevent floods, make use of surface precipitation, and improve aesthetic elements. Some examples of porous pavements are listed below (Winter & Company 2002):

- Porous concrete
- Grass Paving
- Permeable Interlocking Concreate Pavers such as "Store Eco"
- Reinforced gravel such as Gravel Pave

(14) It is recommended that only one type of special pavement material (with the ability to apply diversity in its design and arrangement patterns) be used throughout the sidewalk network. This special pavement can be used along the main streets of the city center, over the width of the sidewalk (from storefronts to the curbs), or within the amenities area (such as bus stops) in a diverse but at the same time integrated way.

(15) Cement pavements are recommended for secondary roads in the city center (Paumier 2004).

(16) Pavement materials should be compatible with the climatic conditions of the target area.

(17) The pathway pavements should help ensure safety, and, in the same vein, making unnecessary changes to them that are dangerous for pedestrians must be avoided.

7.2.3 *Planting and Trees*

Design guidelines:

(18) Street trees should be planted along the sidewalk edge to help define a specific area for pedestrians.

(19) Street trees should be planted at the same distance from the curb and at the same distance apart.

(20) Planting shady trees, bushes, and surface vegetation is recommended to achieve the following goals:

- Smoothing the border between functional areas and different uses;
- Improving the definition of street space;
- Protecting pedestrians from weather conditions;
- Improving the micro-climatic conditions of the area by adjusting temperature, humidity, etc.; and
- Contributing to filtration and natural drainage by reducing the speed of surface runoff and increasing the amount of water absorption in the area.

(21) The design of planting areas should be in harmony with other furniture.

(22) Wherever possible, planting areas should be combined with other furniture. The appropriate combination of trees, plants, and other landscaping elements can be used to create a safe and comfortable environment for pedestrians and a place to rest.

(23) Green medians and pedestrian refuge islands are suggested to slow down cars, make it safer for people to walk, and soften the space on the street.

(24) Taking factors such as the type of trees, the amount of sunlight required, and the size of tree umbrellas and their deciduousness into account, careful planning should be done regarding planting used in the streetscape of the city center.
(25) The trees on the pathways should not reduce the visibility of store facades.

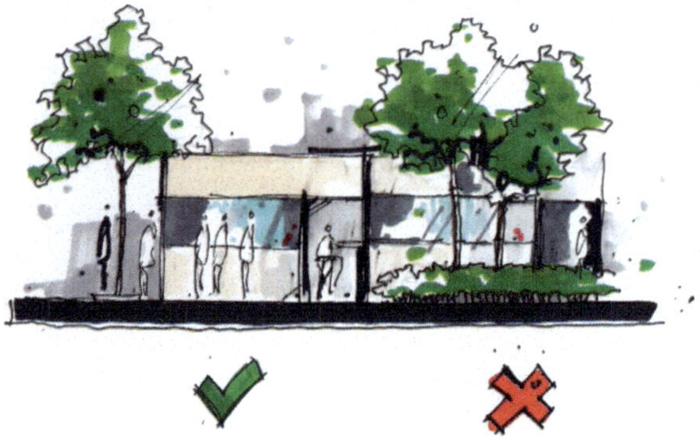

(26) The use of climbing plants in city center roads is not recommended, except on empty walls and partitions, and when it is the only way to create green surfaces on utility lines (LMN 2005b).
(27) Irrigation and maintenance of trees and plants should be possible without disturbance.
(28) Seasonal changes and the gradual growth of plants or trees should be taken into account in selecting and establishing trees and plants. For example, it is necessary to maintain an appropriate distance between two trees so that their crowns do not interfere during maturity.

(29) The trees should be planted at a distance of 12–15 m from intersections and street corners (Winter & Company 2003).

(30) Branches that protrude into the edge of the green medians and pedestrian refuge islands must be pruned at a height of 4 m above the pavement in order to prevent collisions with tall cars (DMAP 2004).

(31) Using planters and tree areas with protective grids can improve the visual quality and streetscape and help direct and drain surface water.

(32) Narrow planters and tree areas with a maximum width of 1 m along the sidewalk and special flowerboxes in front of windows are recommended to enhance street life (Community Development Department 2003).
(33) Portable flowers and plants can be used at regular intervals within the curb limits of sidewalks, window arches, etc. to improve the visual quality of the streetscape. In addition, large removable planters are preferred over fixed gardens because of their ease of maintenance as well as flexibility in the location.

(34) Trees and plants should fit the climate of the design area and, if possible, should include species resistant to drought or extreme cold.

(35) Planting deciduous and umbrella trees are recommended in city centers.

(36) Trees should not obstruct the view at intersections.

(37) Narrow gardens along the sidewalk should not form a continuous strip, as this could impede movement between the roadway and sidewalks and reduce the visibility of pedestrians to drivers and vice versa.

(38) It is recommended to use different kinds of plants to soften and create vitality in ugly and unpleasant walls.
(39) The surface of the tree pit should be leveled with the sidewalk using tree grids.
(40) When using tree grids, care should be taken so that the grids:

- never prevent the growth of trees; and
- are designed in such a way that allow for the installation of upward lighting.

7.2.4 *Street Furniture*

Design guidelines:

(41) The distribution and arrangement of furniture in the open spaces of the city center streets should be appropriate and well-planned.

(42) Street furniture should be placed in areas with high pedestrian activity. Therefore, it is suggested to place furniture at the intersection of pedestrian crossings, the main building entrance, and near outdoor gathering places.

(43) Street furniture should be grouped if possible. For example, gardens and trash cans can be used to shape the space for benches.

(44) All main street furniture should have similar and homogeneous materials and colors.

- The proposed new furniture should be compatible with the existing pieces.
- Single furniture elements should be designed in a way that they can be combined with other ones.

(45) The accumulation of urban furniture, signs, urban facilities and the like should not clutter the street view.

(46) Outdoor dining furniture should match the existing furniture and complement the street view.

(47) Obstacles should be created to smoothly separate pedestrian activity areas from eating places.

- Gardens, fences, chains, and vertical poles are suitable components and elements for this purpose.
- These obstacles must be visible and removable.

(A) **Sitting areas (seats)**

Design guidelines:

(48) Pedestrian sitting areas can be created using edges and steps, low walls, removable tables and chairs, and common benches.
(49) Low walls or garden edges that are used for sitting should be at least 30 cm wide in order to be more effective to this end.
(50) Pedestrian seating areas should be placed in high-traffic areas or places of interest to people.

(51) Benches should be placed in locations where they can provide a sense of comfort and convenience to pedestrians.

- Avoid placing benches along street curbs.
- A buffer should be created between the bench and the street.

(52) The design of the benches and the seating areas should be appropriate and coordinated with the location of possible user groups, their desired interaction and communication, project visions, etc.

(B) **Kiosks**

Design guidelines:

(53) Kiosks should be used for commercial purposes or to give information to the public.
(54) Kiosks should be placed where they are compatible with other street furniture, such as lights and benches, combined with and integrated into their surroundings.

(55) Kiosks should not reduce the visibility of cars or passers-by or block the view of commercial signs.

(56) Kiosks such as newsstands, information booths, etc. should be placed in high-traffic areas or places where pedestrians gather along the main street and should not be scattered and disorderly.

(57) The kiosks must be accessible, visible from all directions, and have sufficient lighting.

(58) Kiosks may not necessarily be fixed and can be made mobile so that they can be moved flexibly in public places.

(59) Public transportation stations should be located close to passenger attractions (cinemas, theaters, department stores, etc.).

(60) Public transportation vehicle stations should be close to main pedestrian crossings and intersections.

(C) **Garbage cans**

Design guidelines:

(61) Garbage bins should be in harmony with other street furniture and be in proportion with the environment. However, they should be characterized by a complementary style and distinctive color.
(62) In order for pedestrians and municipal workers to positioned in high-traffic areas, significant gathering spots, and key locations.
(63) Garbage cans should be fixed most of the time in order to reduce the risk of vandalism.
(64) The bottom of the garbage cans must be closed, and they should have lids so that the contents are kept away from the eyes of passers-by.
(65) Restaurants or coffee shops that have outdoor seating areas should place additional garbage cans near their seats.

(D) **Bike racks**

Design guidelines:

(66) Bike racks should be installed in readable and visible places, near the entrance of commercial places, in an area equal to at least 10% of the space dedicated to car parking (Gould Evans Goodman Associates 2005).

(67) In order to encourage the use of bicycles, reduce the traffic load, and thus increase the functional quality of the main street, bike racks should be installed permanently and be fixed at accessible points throughout the main street public space.

(68) Bike racks should not block the view and do not pose a danger to pedestrians and drivers.

(69) Bike racks should have a simple yet attractive design and allow easy and safe use by the public.

(E) **Facilities and equipment**

Design guidelines:

(70) Facilities and equipment should be safe and easily accessible and not hinder the movement of pedestrians.

(71) The equipment must be combined with and improve the quality of the surrounding environment.

(72) The manholes should be at the same level as the pavement (on both the roadway and the sidewalk) so as not to impede the movement of pedestrians and riders.

7.2.5 Lighting

Objectives:

- providing sufficient lighting to ensure safety and security;
- encouraging and improving nighttime activities;
- upgrading and improving the visibility of buildings and special spaces through lighting and creating a special character for them.

Design guidelines:

(73) It is necessary to provide lighting for the sidewalk surface and building entrances.

(74) The lighting of buildings should be provided and implemented in the same form as at higher levels, and the sidewalk lighting should be provided and implemented at a human scale in the pedestrian areas. Commercial building owners can also be encouraged to keep ground floor lights on during the night.

(75) The placement of streetlights should clearly show changes in road conditions, traffic boundaries, the widening or narrowing of the road, and intersections.
(76) Creating accent lighting in architectural elements, specific landscapes, or historically and historically significant buildings should be encouraged in order to create attraction and focal points.

(77) Placing private lamp posts by restaurants, cafes, ice cream shops, etc. along the sidewalk is prohibited unless they are similar to or coordinated with the entire sequence.

(78) The color of the lighting should be considered in the city center lighting design. For example, metal halide can be used for public use outside buildings, parking

lots, and sidewalks, as well as for street lights and street tree lighting. The use of low-pressure sodium lamps that emit yellow light is not recommended (LMN 2005b).

(79) Uplighting trees and seasonal lighting are encouraged.

(80) Short post lights (minimum 30 cm) can be used along all sidewalks inside the commercial district of the city center (LMN 2005b).

(81) Pedestrian-oriented lighting is required along all streets and in all public places and areas. Fixed pedestrian-oriented lights are between 3.5 and 4.5 m high and complement the general building architecture in terms of character (Paumier 2004).

(82) The city center and the surrounding neighborhoods should take every precaution to reduce light halos and glare.

(83) The height of parking lot lights should not exceed 7.5 m, and their glare should be minimized (Paumier 2004).
(84) In historical city centers, lighting design should increase people's ability to perceive the historical character of the streets and not overshadow it (Winter & Company 2002).

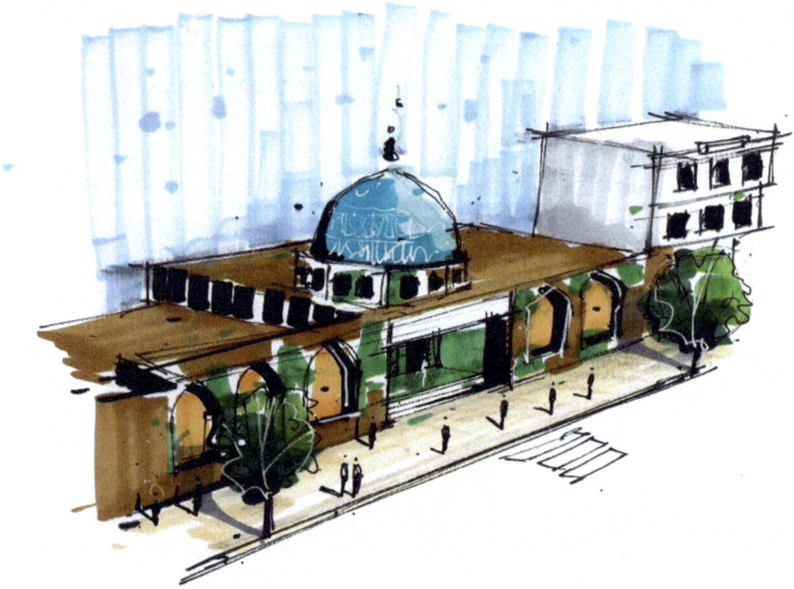

(85) Sidewalk lighting should be at least 5 lx (Pakzad 2013).
(86) The lighting of intersections should be equal to the sum of the predicted lighting values of each street. This lighting, however, should not be brighter than 10 lx (Pakzad 2013).
(87) The branches and leaves of trees along the street should not interfere with the lighting required for the street space.

(88) Hidden corners on the street should have at least 50 lx of lighting at night (Pakzad 2013).

(89) Floodlights should have half-transparent lenses made of frosted glass or acrylic to cut down on glare.

(90) In order to achieve aesthetic goals, streetlights should not only fulfill their functional purpose, i.e., providing lighting for the thoroughfare, but also enhance the visual appeal and create a sense of human scale.

(91) The lighting sources on the roads should be waterproof, and their light should be covered in such a way that it does not disturb the vision of passers-by or drivers.

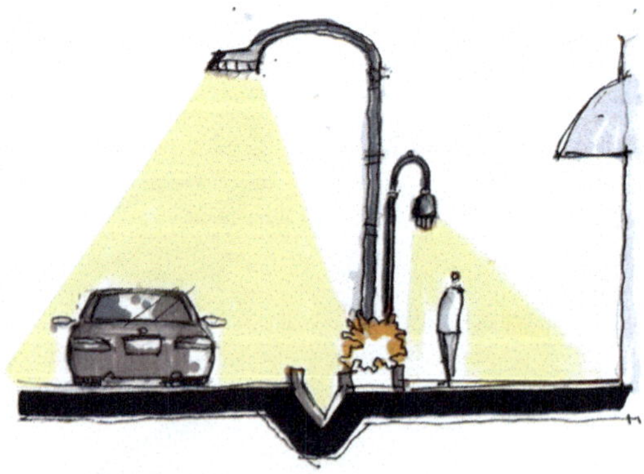

(92) Canopy lighting should illuminate only the area under the cover.

(93) The lighting of the main axis in the city center should not disturb residential units located on the street. Also, this lighting should not be dazzling and must not penetrate inside buildings.

7.2.6 Fountains and Water Features

Design guidelines:

(94) Fountains and water features should be located in key public places, such as plazas or on main street corners, where they can attract more people.

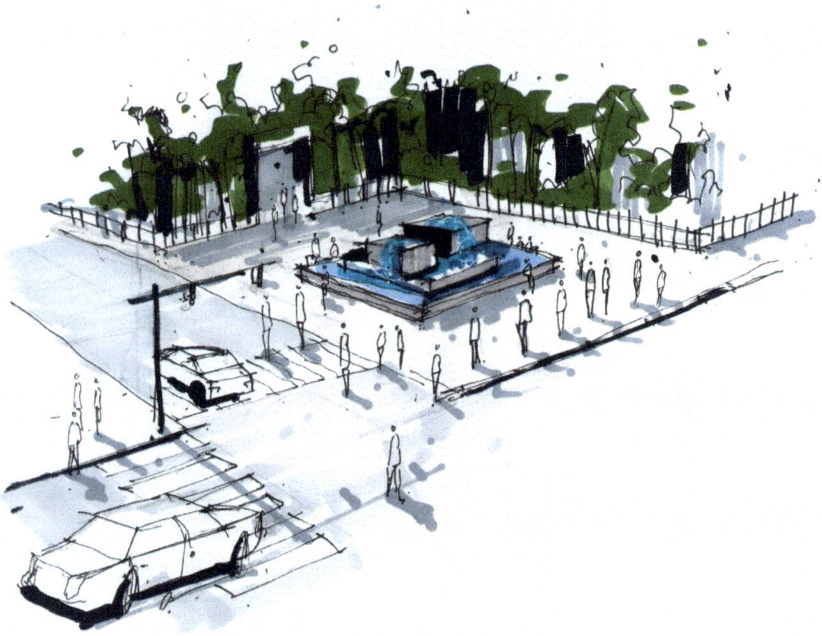

(95) There should be maximum accessibility for pedestrians in all directions around water features.

(96) Using arrangements such as musical fountains, which combine water features with sound and light, can encourage more people to sit on the edges and engage in social interaction with each other.

7.2.7 *Public Art*

Objectives:

In order to protect the city center, public art should contribute to humanizing the environment.

Design guidelines:

(97) Using collective works of art in city centers and main streets are highly recommended.
(98) Places such as entrances, pedestrian bridges, small plazas, and terminals are among the primary target points for installing public artworks.

(99) Public art must be related to the cultural, social, and historical background of the city center and the entire city or express a story or an event for the residents. If possible, they should be made by a local producer (artist or craftsman).

(100) Public artworks will be more desirable and valuable if they are integrated into the street architecture or landscape and considered an integral part of it.
(101) These works must be easily accessible and highly visible, but they should not impede pedestrian or vehicular traffic.

(102) The artistic design of benches, tree grids, manhole covers, fences, canopies, signs, etc. should create remarkable and commendable artistic components.
(103) They should be made of durable and weather-resistant materials.
(104) They should have artistic subtleties and avoid an unpleasant and rough appearance.

(105) These works can include water and fountains, seating, green spaces, architectural decorative elements, or plazas, as part of their design elements.

(106) Their design should eliminate the possibility of risks and physical injuries to people who may come into contact with them.
(107) The use of vertical elements in special places for placing works of art should be avoided.
(108) Art programs should be developed for the city center main street, with which the design and placement of any collective artworks can be coordinated.

7.2.8 Public Spaces

Objectives:

- Public spaces in the city center should act as a strong visual focus as well as an important activity center.
- Public spaces aim at providing pleasant, desirable places with beautiful landscaping for the gathering of people.

Design guidelines:

(109) Public spaces in the city center should be designed to maximize their accessibility, with an emphasis on their placement in the gravity center of the area (McIlroy 2006). The primary building entrances should face public open spaces, and the passageways should be designed to facilitate movement.

(110) Public spaces should have the ability to adapt to a wide range of different functions and connect planned activity edges.

(111) These spaces should provide a specific and diverse number of seating places and use the best possible quality design and materials.

(112) Public spaces must be incorporated into the integrated sidewalk system (Paumier 2004). In other words, city-center public open spaces should emphasize the continuity of the main street front. In general, the dominant portion of a block's edge should consist of building walls. The gaps created in the main street wall for the establishment of public open spaces should not overshadow the continuous and defined edge of the street.

(113) These spaces should create a positive visual structure for the city center as well as a specific local identity for adjacent constructions.

(114) It is recommended to use distinctive edges for public open spaces along the sidewalk. To this end, changes in pavements, edges, and walls can be used to define the edge of the street.

(115) Public open spaces in the city center should be established in such a way as to increase the possibility of benefiting from sunlight (in cold weather) and shade (in hot weather).

Sunshine in the cold season

Shadow in hot season

(116) The creation of construction projects with introverted functions and activity in the city center should be avoided unless these complexes consider their introverted arena as a public space with a specific use in relation to the street axis.

(117) Public spaces should be designed in such a way as to provide unobstructed views. Also, lighting as well as dynamic and lively activities in the late hours of the night should be provided to increase their sense of safety and security.

(118) Important public spaces or main signs should mark intersections in the heart of the city center.

(119) A public space can be designed and constructed as a shared area between multiple buildings or as a separate public space on a particular site.

(120) Public open spaces should be properly considered in one or more of the following forms (Lenexa 2007):

Plaza: A plaza is an open, paved space that is typically created by a pedestrian-oriented space, trees and plants, seating areas, and occasionally a few retail shops. In the city center, this type of open public space can be used for resting, stopping, and large gatherings of people. It is suggested that plazas be constructed at public transportation stations and other key locations. They can provide the necessary space for outdoor exhibitions and retail and entice people to visit.

Green space: These spaces are covered with grass and surrounded by trees, plants, and sidewalks. Such spaces can be used for large gatherings and recreation.

Courtyards: Courtyards are partially or entirely enclosed play areas surrounded by buildings. However, they are less useful for large gatherings and can be considered semi-private.

Forecourt: Forecourts are open areas created in front of buildings, which may contain benches, short walls, or tables and chairs. These areas may be owned privately or publicly.

7.2.9 Important Entrances to City Centers

Objectives:

Defining and specifying important and main entrances to the city center.

Design guidelines:

(121) Constructions along main entrances into the city center should not be uniform in terms of usage, scale, and importance (Paumier 2004).

(122) On the main entrances into the city center, buildings should be made impressive and exciting by architectural measures including the following (LMN 2005a):

- Tower forms at corners
- Landscaped forecourts
- Large entries
- Deep cornice overhangs for buildings at corners
- More intense colors
- Impressive lighting
- Special and unusual shapes and forms
- Artworks such as sculptures or murals

(123) The city center entrances and their facilities and amenities may also be combined with a public space to serve both types of functions and give such entrances a more prominent role.

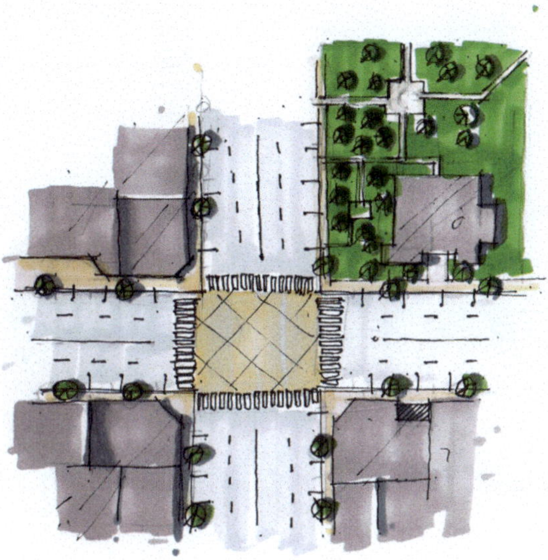

(124) The entrances should be designed in such a way that they are recognized as high-quality public spaces for people and visitors. Below are examples of entrance designs for city centers.

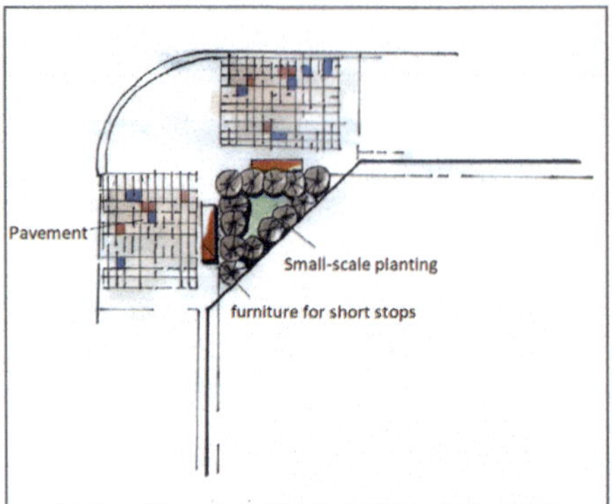

Creating entrances at corners with small landscaped forespaces

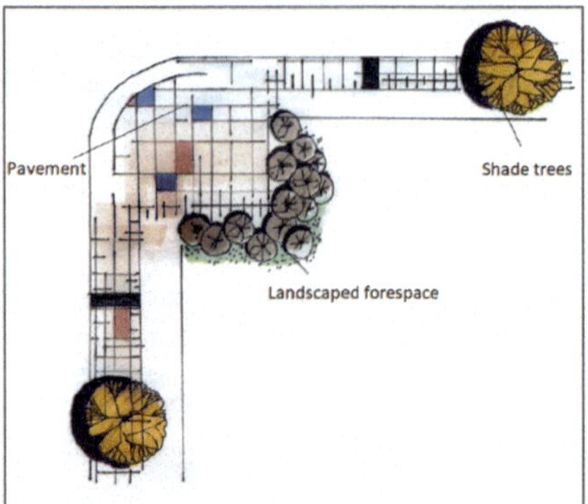

Creating entrances at corners with small landscaped forespaces

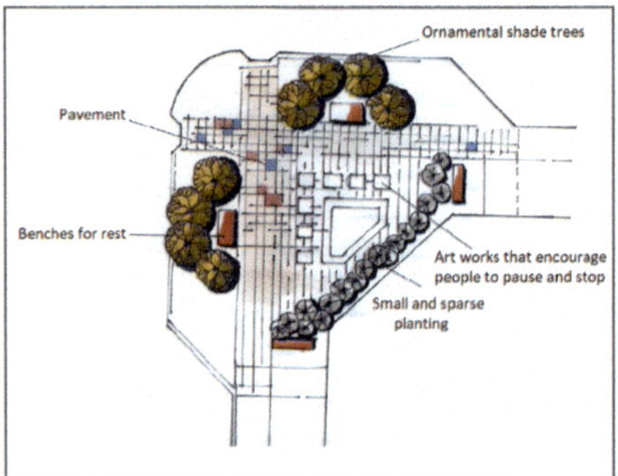

Creating entrances at corners with widened turns and small-scale plaza

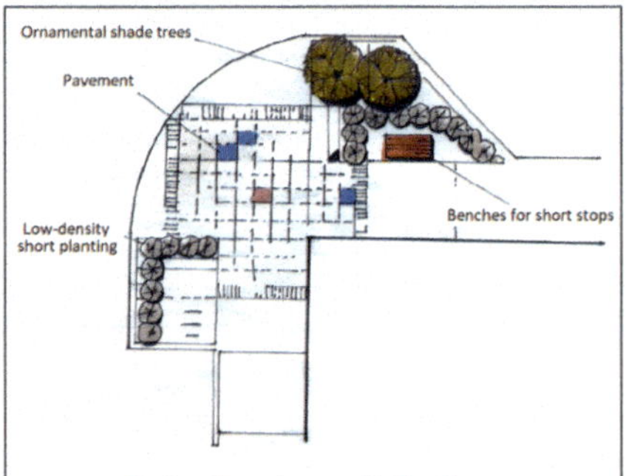

Creating entrances at corners with widened turns

7.2.10 Open Spaces

Objectives:

Creating various open spaces in connection with each building helps to improve the visual continuity of the city center and creates more options for gathering, meeting with friends and acquaintances, and enjoying the fresh air.

Design guidelines:

(125) Every new building must allocate at least one percent of the total building area and site area to open spaces (LMN 2005a).

(126) Such spaces should be created where they can be seen and accessed from sidewalks and other pedestrian passages.

(127) These open spaces should include at least some of the following facilities (LMN 2005b)

- Seating furniture (benches, ledges with a height of 38–45 cm, etc.).
- Planting, including trees, shrubs, and seasonal plants
- Pedestrian-scaled lighting
- Artworks
- Water features
- Decorative pavements
- Canopies
- Grids for supporting and protecting trees
- Other approved facilities in line with the purpose of this guideline.

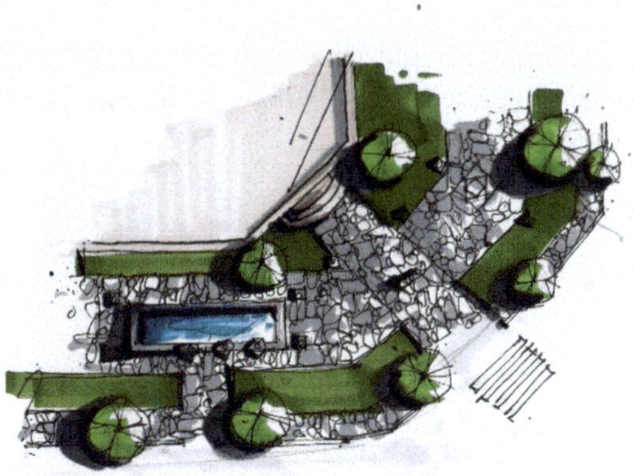

7.3 Design Guidelines for the Architecture of Buildings in City Centers

7.3.1 Creating a Unique Character for Buildings in City Centers

Objectives:

Creating groups of buildings that present a single character and, at the same time, provide opportunities to show the architecture of individual buildings.

Design guidelines:

(128) The design character of a single building should fit that of the surrounding buildings (i.e., they should have common features), and at the same time, they should include unique and distinctive features. In this regard, applicants for construction in the city center (especially for buildings facing the street) should first present their architectural plan to determine the similarities and differences their proposed building will have as compared to the context.

Similarities and differences are related to the following factors:

- Building shape and density;
- Roofline and appearance of shelters;
- External building additions such as signs, artwork, special canopies, entrance landscaping, and so on;
- Window arrangement, size, orientation, and placement;
- Materials and colors; and
- The way a building is connected to the site.

Finally, a special commission related to the design of city centers should approve or reject such proposed plans.

(129) Residential buildings should be designed to complement the surrounding commercial buildings and have an urban character rather than a suburban one (LMN 2005b).

(130) New interpretations of traditional building styles should be encouraged in the city center. A new design is preferred that is based on basic similarities between old buildings in the city center but does not copy them (Winter & Company 2002).

(131) Pure imitation of old building styles in the city center is not recommended.

(132) Wherever possible, examples of traditional architecture in commercial and residential buildings in the city center should be protected, renovated, and adapted in proper places to accept new uses.

(133) The concentration of retail shops should increase near the central core of the city to create a special visual character there (McBee et al. 1992).

(134) A correct visual indication should be created for movement toward the city center by reducing the spaces between buildings and increasing their height (increasing density).

(135) As much as possible, the creation of urban superblocks (with a length of more than 90 m) should be avoided in the city center (Paumier 2004).

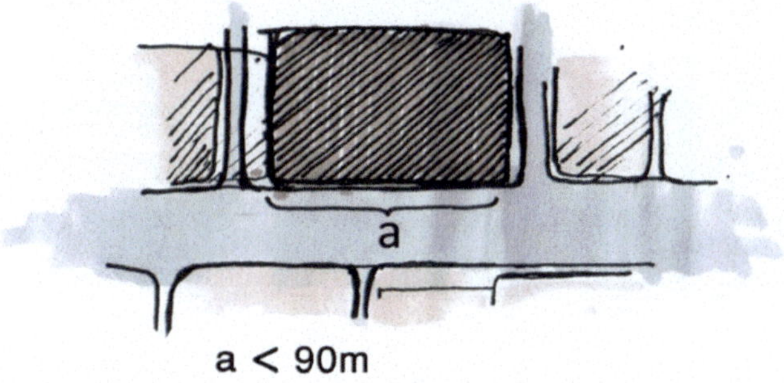

(136) In a historical city center, when approaching historical buildings, the new buildings should be shorter so that they do not reduce the visual value of special historical buildings.

(137) Historical materials of facades, decorative borders around doors, and historical cornices and gutters are among the defining features of the character of city center buildings and should be preserved.

(138) The defining features of historical buildings in city centers must be preserved. They should not be lost or damaged and should be replaced if damaged.

(139) Changes in old buildings in the city center should be in accordance with their historical character. Moreover, changes that hinder the interpretation of the historical significance of the building or induce a time period prior to the building's creation should be avoided (Winter & Company 2002).

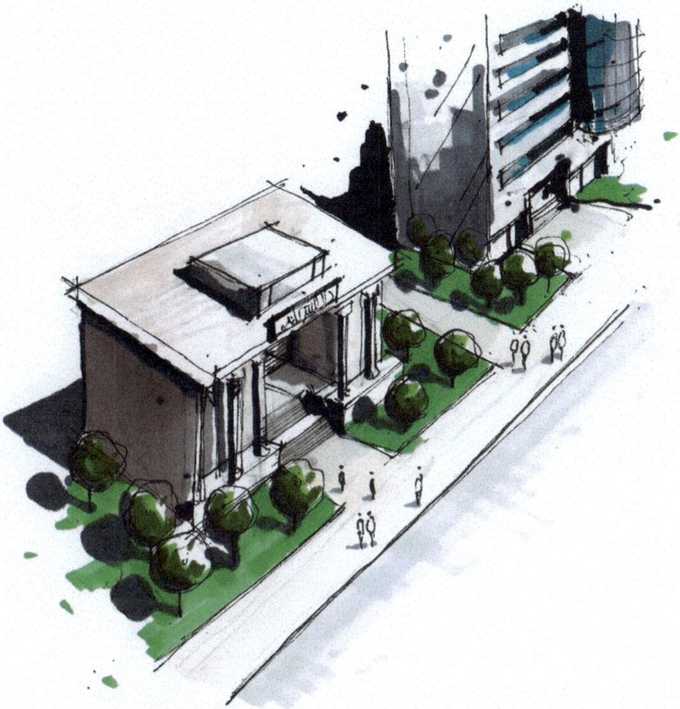

(140) Local or regional architecture should be encouraged instead of foreign architecture.

(141) Building floors (except in the parts where there is a possibility of surface water runoff after raining) should not be lower than the ground level or higher than 6.0 m above the ground level (LDR 1998).

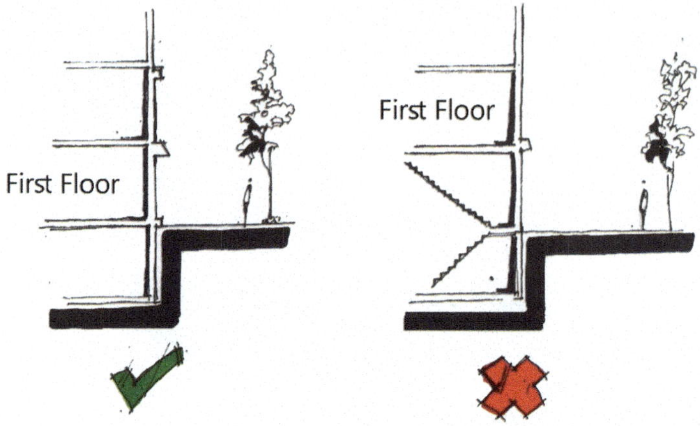

7.3.2 Design Integrity

Objectives:

Creating unity in the design of city center buildings and forming an integrated environment with continuity and rich visual value.

Design guidelines:

(142) An integrated architectural scale in buildings contributes to an integrated design in the entire city center. This scale should be such that a more focused character is created near plazas, public places, and spaces along the city center's main commercial streets.

(143) In historical and old city centers if a new building is placed in the city center, it should observe the rhythm line of horizontal components (such as cornices, strips, plinths, lines of shop fronts, lines of windows, etc.) at the edge of the block.

(144) If we have to create a taller building in the city center, the higher floors should be set back from the main facade, and the lower floors should be placed along the edge created by existing buildings.

(145) New developments in empty gaps should create a balance between old and new styles.
(146) Where existing visual richness and design standards are poor, repeating their design components will not be desirable, and new buildings should be used to formulate new contexts and standards (Paumier 2004).

(147) The height of city center buildings can vary, but they should have a strong sense of similarity in scale.
(148) The smaller the city center area and the fewer buildings there, the greater the importance of building integrity and the less accepted contrasts in forms, materials, volumes, and building heights (Mcllroy 2006).
(149) Using constructions to fill empty spaces in the urban context, especially on the main arteries of the city center, and creating continuous edges along the streets, is recommended.
(150) Coordinated fundamental decisions should be made regarding setbacks, the height of buildings, the overall organization of views, etc., along the main street of the city center, and measures should be taken to create visual continuity through similarities in building height, scale, massing, the overall organization of views, materials, colors, and roof shape.

7.3.3 Articulation of Buildings

Objectives: Preventing long and boring façades.

Design guidelines:

(A) **Linear separation (horizontal)**

(151) In order to avoid creating long and uniform facades in the city center, long buildings (e.g., over 20 m in length) parallel to the roads or public open spaces should be distinguished, i.e., articulated, by regular intervals. The following

guidelines are recommended for building articulation (LMN 2005b; Lenexa 2007):

- Modulation via setbacks or protrusions of part of the façade;
- The use of special building components, such as balconies, arcades, canopies, towers, and entrances, to visually divide the facade into smaller parts.
- Use of focal points in buildings, including distinctive entry features, etc.;
  - Changing roof lines;
  - Changing materials;
  - Other approved methods.

(B) **Vertical modulation**

(152) Tall buildings should also be divided into smaller parts on the facade. This vertical separation should be applied to building fronts parallel to roadways or public open spaces in the city center.

(153) If a large building is divided into smaller parts, this separation should be emphasized three-dimensionally throughout the building (LMN 2005b).

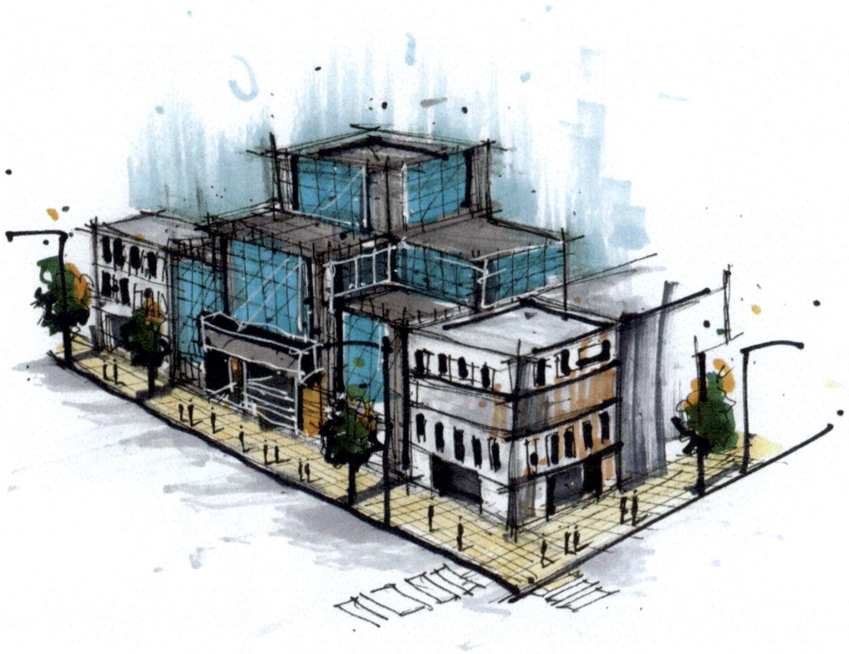

7.3.4 *The Connection Between Buildings and Sidewalks*

Objectives:

To ensure that city center buildings bring life to streets and sidewalks.

Design guidelines:

(154) City center buildings should be connected to each other through landscaped sidewalks, courtyards, small-scale parks, walkways, and public spaces.

(155) Within the city center, buildings should be oriented toward the main streets, not toward an open parking lot or parking structure. In general, if buildings are not tangential to sidewalks, they should be within their perimeter. Facades adjacent to sidewalks should include windows, entrances, canopies, and other elements (LMN 2005b).

(156) Including some areas for expansion of uses in the sidewalk, such as the space under porches or canopies, as well as the use of building roofs, can increase the dynamics of the street space.

(157) Buildings can be set back from the sidewalks to create a public space or a usable front for pedestrians, as well as green cover and landscaping for the roadways.

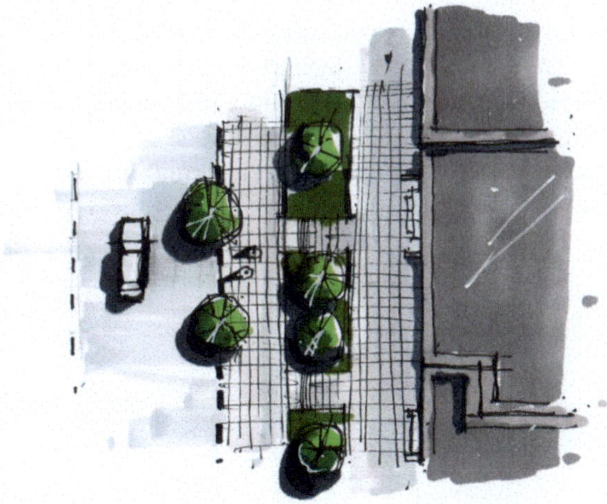

(158) The presence of commercial uses (retail shops) on the ground floor level and special uses that attract the crowd (such as a theater, cinema, or gathering center) is a must in the city center's main streets.

(159) Multi-story buildings with commercial, office, and residential uses are highly recommended as they can create clear and defined edges for the street and bring about activities on the upper floors that overlook the street and public spaces in the city center. In the meantime, easy access to the upper floors should be provided to ensure the continuity of commercial uses and separate entrances for residential uses.

(160) Destruction of valuable buildings (from a formal, functional, and semantic point of view) at the edge of the city center main street is prohibited.

(161) Safe and comfortable places should be created in the city center's main streets for people to stop for activities such as watching, social interaction, resting, etc. Such places should not interfere with other uses of the sidewalks.

(162) Visual and physical connections should be established among the interior spaces of buildings and adjacent sidewalks (see design guidelines regarding the transparency and prominence of uses).

(163) Infill developments should be used in the city center context to create a continuous edge along the main streets.

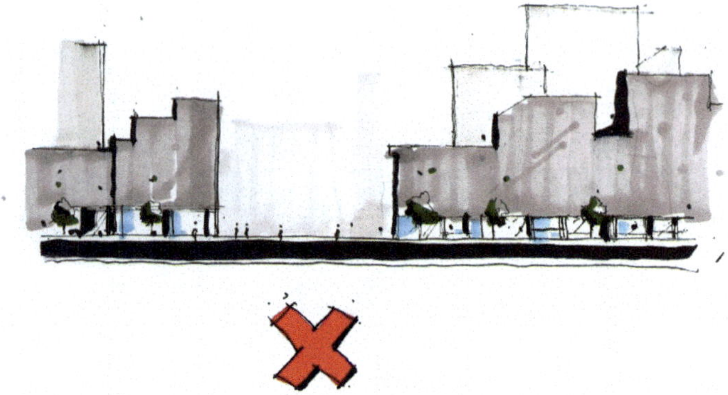

7.3.5 *Building Facades*

Objectives:

Coordinating the design of city center buildings, turning the individual facades of each block into an integrated architectural unit, and strengthening visual continuity.

Design guidelines:

(164) All visible surfaces of architectural bodies (even on the second wall) must have a façade design.

(165) In facade design, a proper connection should be made between the location and ratio of openings in the facade, with a focus on organization and the creation of defined and visible facades on the ground and upper floors.

(166) The presence of monotonous and long facades on the main streets should be avoided. The following measures can be effective in eliminating monotony and ensuring diversity:

- Changing the roofline
- Changing the level of the walls
- Indented doors and windows
- Covered sidewalks, arcades, and canopies
- Other architectural details in harmony with the context.

(167) Creating strong horizontal connections in the facades of the city center's main streets through the continuous repetition of display windows, a coherent frieze, and colored canopies connecting buildings, etc.

(168) Buildings should show distinctive points in their facades in transition from the ground floor to the upper floors.

(169) If valuable parts, elements, and details of old building facades in the city center have remained intact, they should be preserved, but if they have changed, they should be returned to the original design of the facades. The repetition of such components and elements will undoubtedly create harmony and visual continuity in the city center streets.

(170) Change in materials, projected and indented arches, window frames, hanging furniture, architectural details, etc. can be used to increase the visual richness of the facade on the main street wall (Winter & Company 2003).

(171) Shop facades should be perceived as an open space in the viewer's mind and create an active visual focus that contrasts with the full volumes of the upper facades (Paumier 2004).

(172) The distinction between the lower and upper floors should be visible. Creating transparency in the first-floor view makes the building more inviting. It should also be noted that the use of windows and other architectural elements to create patterns can give building facades rhythm. The upper floors of the building create a distinct pattern due to the greater proportion of filled-to-empty areas and the smaller windows.

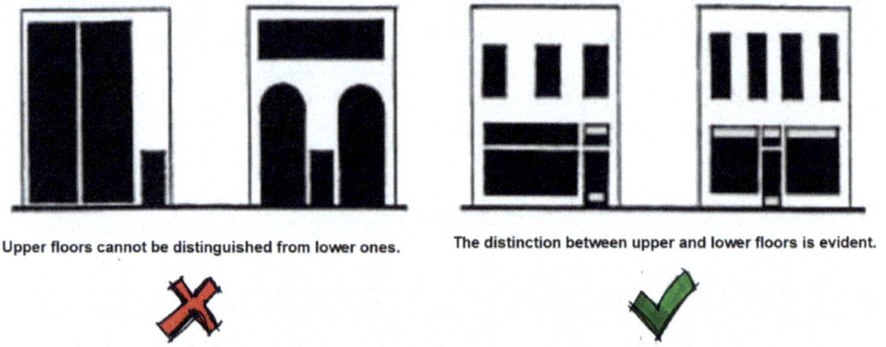

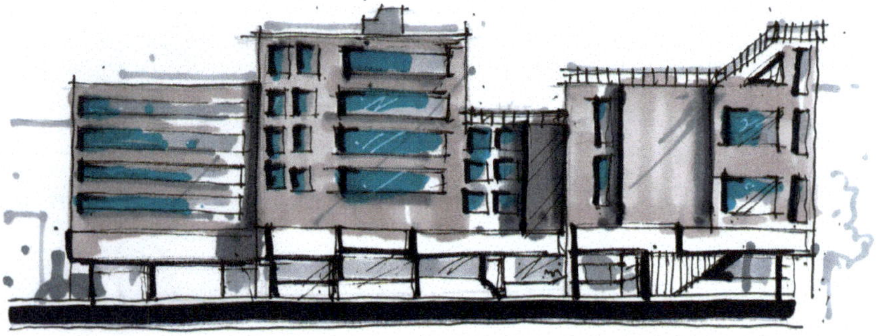

(173) Using details in the shop fronts and the type of doors, windows, openings, etc. embedded therein can add to the visual richness and human scale of buildings.

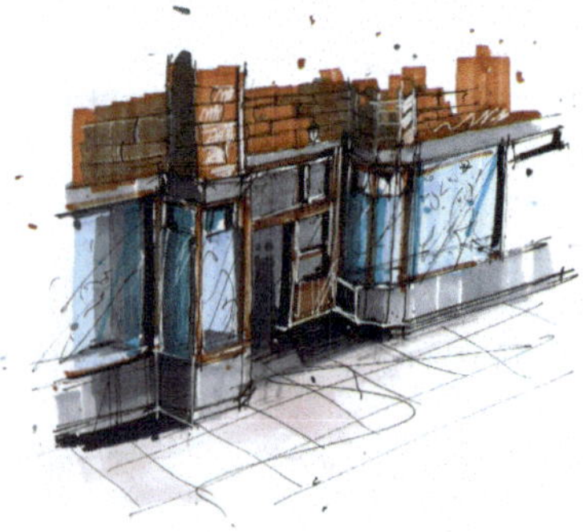

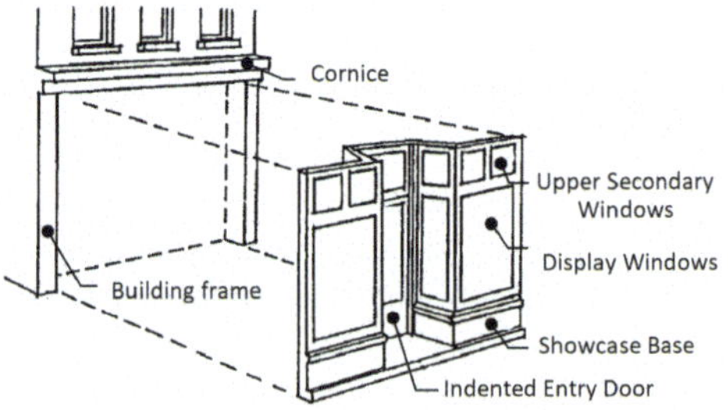

(174) The articulation and breaking up of the building facade (not necessarily large-scale buildings) can be used to achieve a dynamic and attractive building facade.

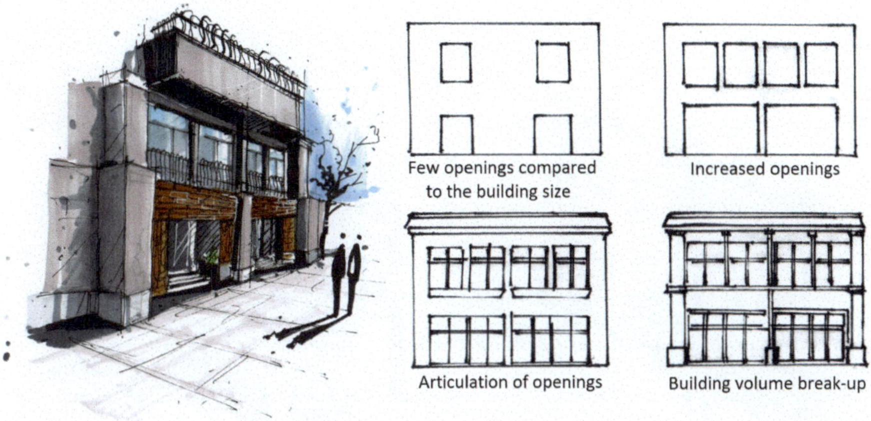

(175) Skyline continuity is required to increase the spatial definability of the street

(176) The walls of lots containing green spaces and landscaping should be as see-through as possible (fences, hedges, etc.).

(177) Though the facades of shops on the ground floor and upper floors should be part of an integrated unity, they should be distinguished from each other via strong horizontal elements that give them their own character.

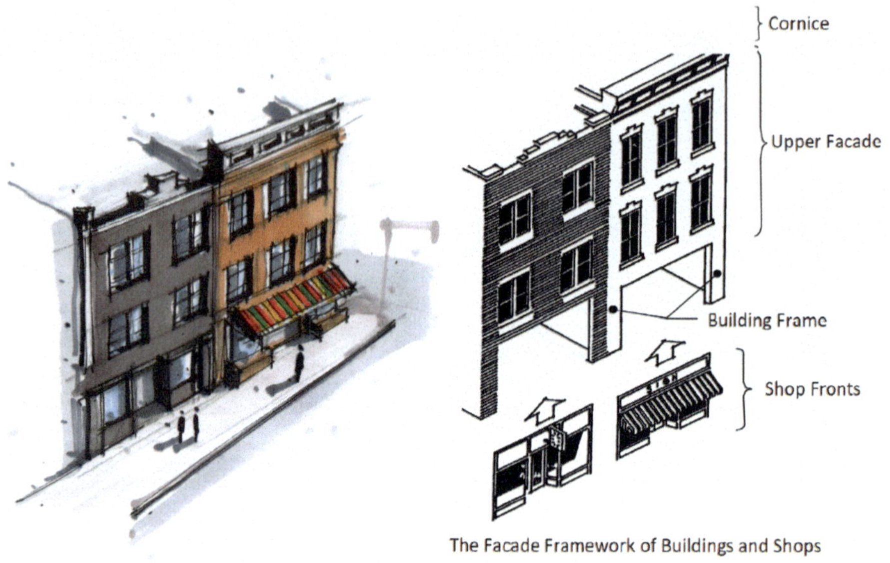

7.3.6 *Building Elements and Details*

Objectives: Increasing the visual appeal, strengthening the rhythm, emphasizing the importance of the sidewalk surface, and creating a human scale for the street.

Design guidelines:

(178) If the cornices, plinths, or other details of the city center buildings have been lost or replaced by something else, they must be restored.

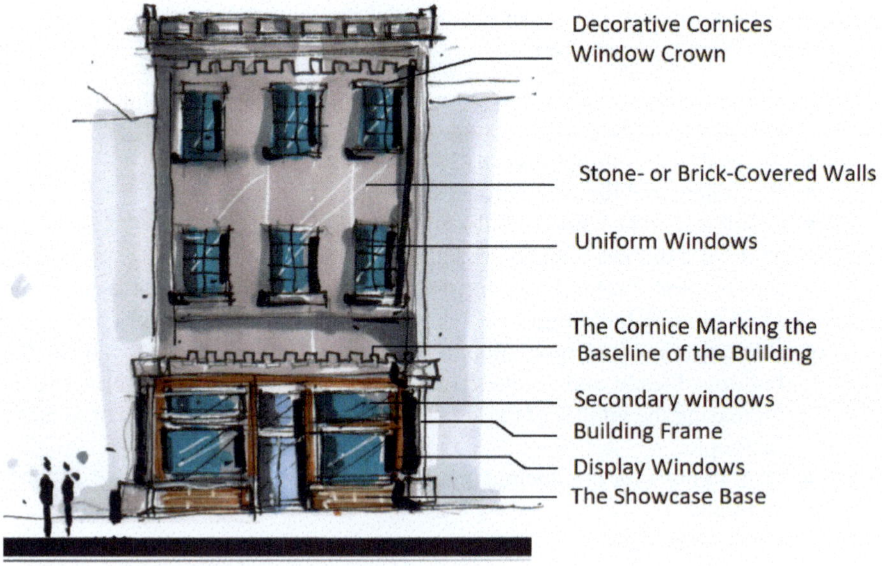

(179) New building details should be designed according to the overall building size.

(180) All exterior surfaces of buildings facing the streets and public spaces in the city center must use the following essential building elements and details to create a pedestrian-friendly scale in the city center: Essential building details and components are those that contribute significantly to the building's form and personality. The "decorative" elements of a city's downtown should exhibit exceptional craftsmanship and a distinctive design to enhance the

visual appeal and distinctive character of the area. These elements include (LMN 2005a; Lenexa 2007):

- Building elements that create joints and rhythm, such as windows, doors, entrances, corners with glazing or geometric shapes, etc.
- Decorative building materials such as tiles and metal works
- Prominent and distinct entrances (covered or indented)
- Canopies, arches, porches, flooring, windowsills, etc.
- Vertical windows with colored glass (multi-piece)
- Fragmented (broken) roof form with vertical separation that reduces the scale of large buildings
- Balconies on the upper floors. Balconies not only help to divide the building facade and create visual appeal, but they are also effective in increasing the vitality of the street image and can create a sense of security in people.
- Wall lights with ornamental brackets
- Artwork
- Hanging baskets
- Medallions
- Belt courses along facades
- Plinths for columns
- Shallow recesses
- Projecting sills
- Special planters
- Pedestrian scale signs
- Other acceptable elements and details in line with the above purpose.

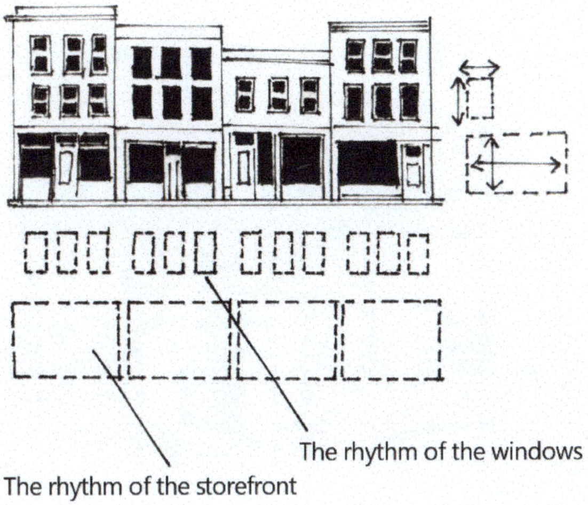

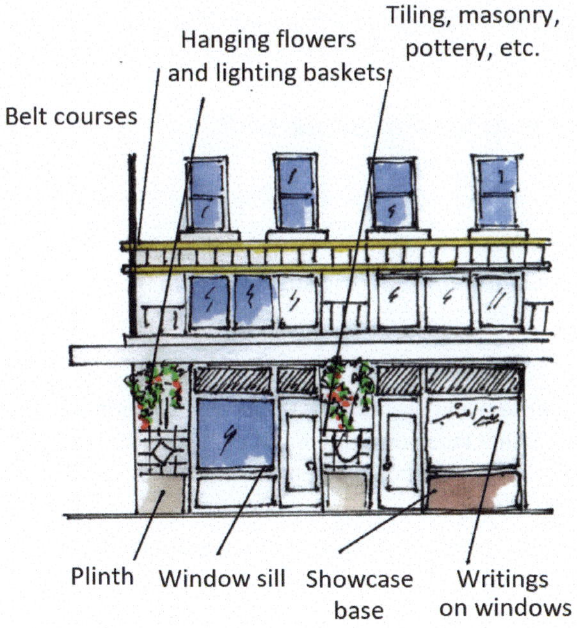

(181) Buildings located at intersections should include special architectural elements and details as well as focal points in the design to emphasize and strengthen the importance of their location (LMN 2005b).

7.3.7 *Provisions Related to Blank Walls*

Objectives:

Special measures should be taken for walls without openings to increase attractiveness and visual continuity and variety in street views in the city center.

Design guidelines:

(182) Blank walls with a length of more than 4.5 m that do not have windows/openings, entries, and architectural elements (modulating and dividing the facade) should not face public open places and main streets of the city center. Where blank walls are inevitable, they should be modified in the following ways (LMN 2005a, b):

- Planters or scaffolds for vines or other climbing plants.
- Landscaping that can cover at least 30% of the wall surface with plants within 3 years.
- Special materials (such as decorative patterned masonry, tiling, etc.).
- Artwork, display windows, opaque or translucent glass,
- Vertical articulation (see building articulation guidelines).
- Installation of lighting tools and equipment, raised metal canopies, raised cornices, flower pots and plinths at the base of the wall (which can be

concrete or stone), continuous belt courses of different texture and color along the wall; and

- Other arrangements approved in line with the said objectives.

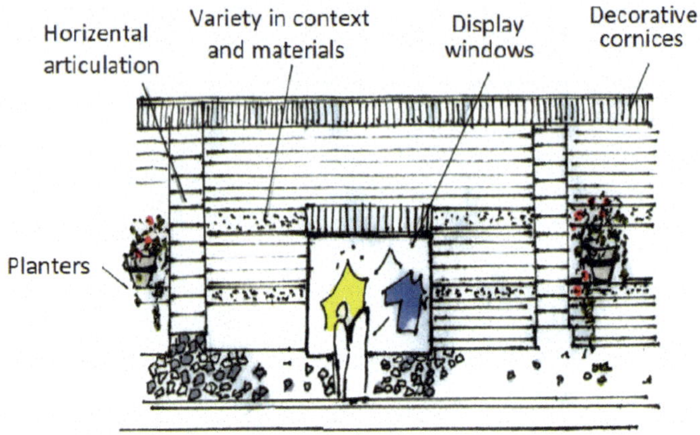

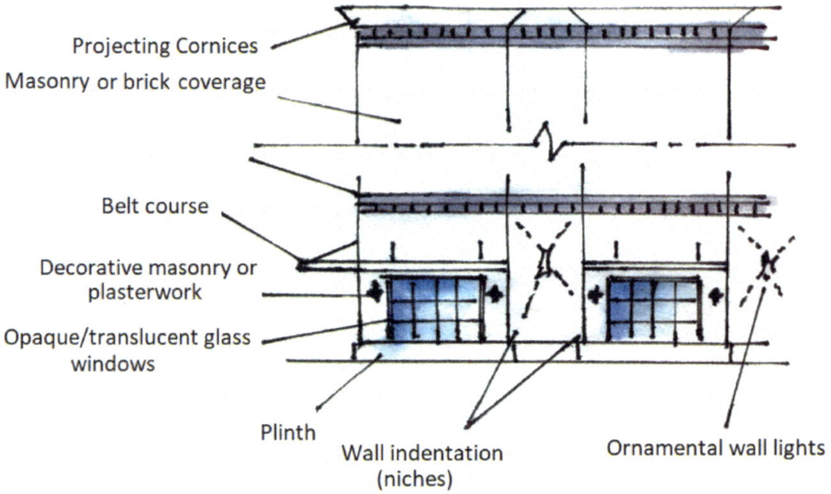

7.3.8 Street Frontage

Objectives:

- Strengthening pedestrian activity and orientation and increasing street vitality through building entrance design.
- Improving environmental security by increasing the number of observers on the street.

Design guidelines:

(183) The entrances should connect the inside and outside of the building according to the type of use and the needs of the users.

(184) Building entrances should be visible from the street and should be oriented toward the sidewalks to provide pedestrians with clear, legible, and easy access.

(185) Entrances should provide easy access to all types of vehicles, including public transportation, private cars, bicycles, and walkways.

(186) Entrances should have a unique architectural appearance that can be created through two or more of the following elements:

- Longer and deeper arches
- Projecting canopies
- Slightly recessed views
- Front porches
- Porches with exclusive roof shapes.

(187) Entries with a little setback should be encouraged because the repetition of recessed entrances creates a pattern of shadows along the street and improves its pedestrian scale. They also provide protection against weather conditions.

(188) The entrances can be set back up to 1.5 m behind the front line of the building (Community Development Department, 2003).

(189) In places where entrances are placed in the same line as display windows, canopies can be used to create diversity.

(190) The glass used in shop entrances should maximize the visibility of indoor activities.

(191) Where buildings have stairs at their entrances, they should not obstruct pedestrian movement or narrow down the sidewalk.

(192) The entrances on city center streets should be made much more prominent than parking lots or side entrances. This can be done through articulation and architectural arrangements.

(193) Where entrances to the upper floors are adjacent to the main shop facades, they should be mixed in with the architectural framework so that they appear as secondary components.

(194) The primary building entrances should be oriented toward the city center streets.
(195) The design style and materials of entrances should be compatible and harmonious with the character and design of the commercial facade.

(196) If entrances are behind the outermost building facade, they should be distinguished and defined using elements such as lights, scaffolding for climbing plants, canopies, and other architectural elements (Winter & Company 2002).

(197) In general, the use of shared doors for residential, public, and commercial uses should be avoided (LMN 2005a).

(198) In historical and traditional city centers, the entrances should be similar to historical ones.

7.3.9 *Transparency and Projection of Uses that Vitalize the City Center*

Objectives:

Establishing uses and activities that attract the population to improve the quality of life in city centers.

Design guidelines:

(199) Visual and physical connections must be established between active spaces inside buildings and adjacent sidewalks.

(200) The ground floor facades of buildings that are oriented toward the city center streets should have pedestrian-oriented uses and transparent windows at an average height of 30–75 cm above the ground (Nottingham City Council 2006) in compliance with the following guidelines (LMN 2005a)

- Along the sidewalks in city centers, at least 60% of the uses must be pedestrian-oriented, attract the population, and have transparency in their main facade.
- Along the main streets in the city center, at least 40% of the uses must be pedestrian-oriented, attract the population, and have transparency in their main facade.
- Along other city center roads (auxiliary, secondary, inter-block, etc.), at least 60% of the uses must be pedestrian-oriented and have transparency in their main facade.

(201) Buildings in the city center can have setbacks up to a certain depth from their property line. Such setback spaces can be used to create building entrances, (outdoor) street cafes, or other activities and uses. In addition, they may be used to include public spaces for pedestrians in the city center.

(202) In order to create high-quality transparency in building facades, reflective, opaque, and translucent glass should not be used in display windows (for a detailed discussion on display windows, see window design guidelines) (LMN 2005a).

(203) The facade of shops should be perceived as an open space in the viewer's mind and create an active visual focus that contrasts with the upper facades (LMN 2005b).

(204) In the main streets of the city center, the number of exposed windows should be maximized on the level that is framed by the plinths and columns of the upper facades (Winter & Company 2002).

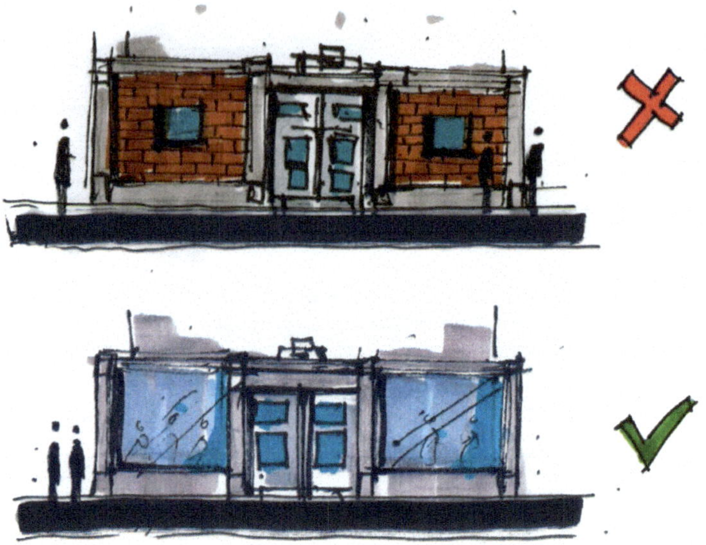

(205) Balconies (with public access) playing the role of hanging gardens facing the city center roads with a view to the pedestrian space are recommended.

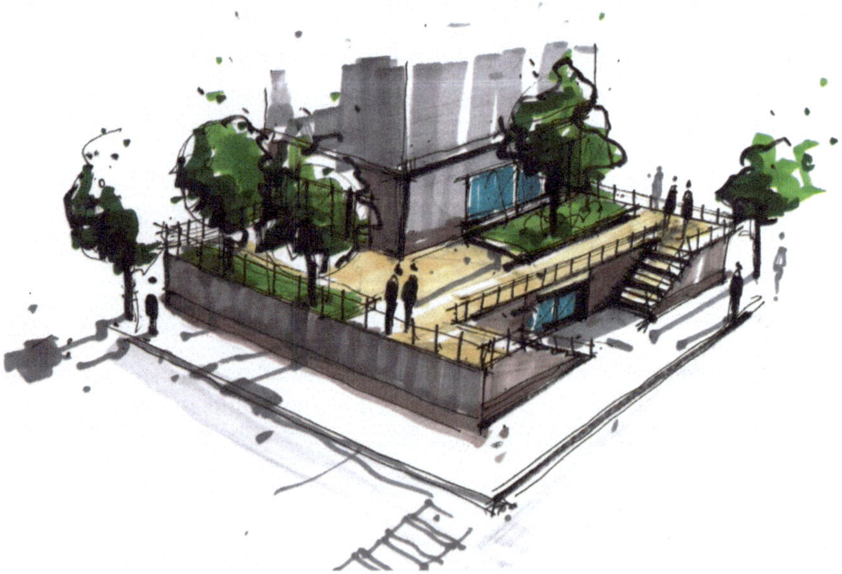

7.3.10 Canopies and Awnings

Objectives:

Creating a canopy for pedestrians against strong sunlight and rain, and creating a pleasant walking experience in rainy conditions in this way.

Design guidelines:

(206) Canopies that provide protection against rain and sunlight should be created along at least 75% of the facade (LMN 2005a).

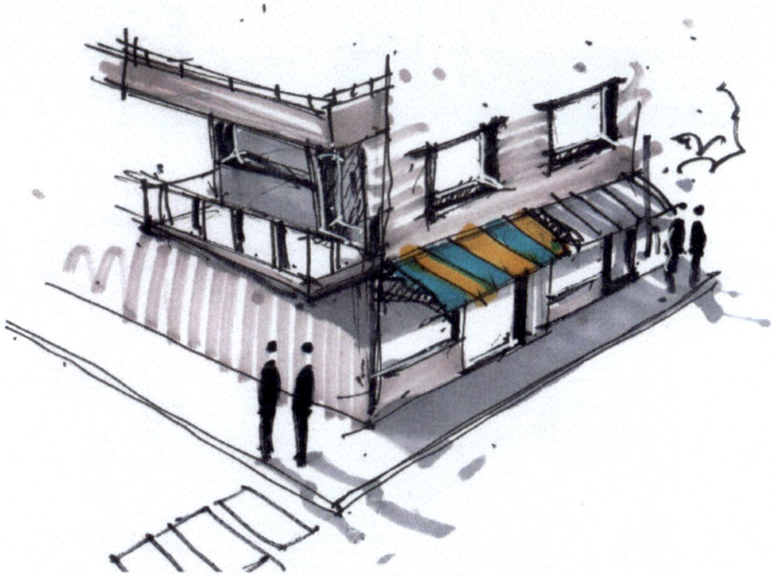

(207) Canopies that provide protection against rain and sunlight should be created along at least 75% of the facade (LMN 2005a).
(208) Canopies can be made of any type of resistant and permanent material, but glass and steel (and other resistant metals) are preferred (LMN 2005a).
(209) Canopies should be directly connected to the building without additional bases and colmns at the sidewalk level.

(211) Hotels, theaters, and other important buildings can use canopies in their architecture to strengthen the entrance and protect pedestrians from weather conditions.

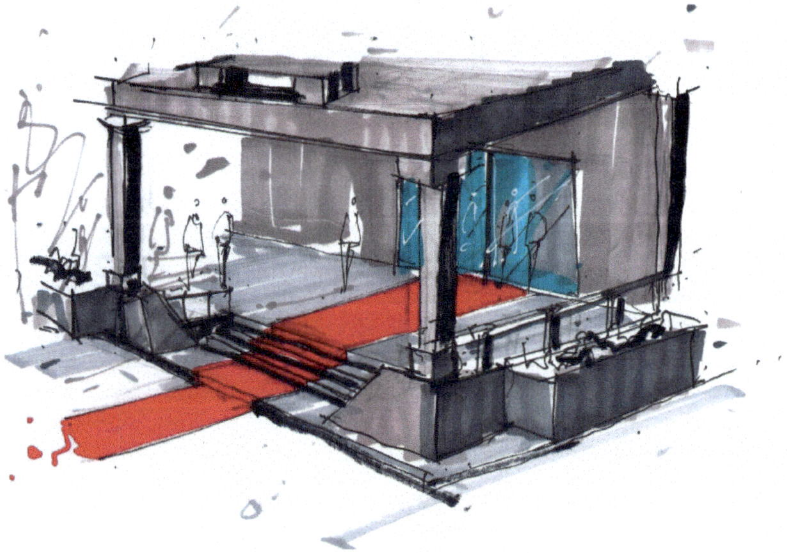

(211) Canopies should be used to focus attention on storefronts and create a strong horizontal element repeated across city center blocks.
(212) Signs can be hung from the canopies (for more information, see the sign design guidelines).

(213) In many cases, it is possible to create protection against rain and sunlight through the use of arcades, setbacks, projections, and other measures.

(214) The color, design, and size of canopies should match the building facades.
(215) The use of removable or changeable canopies is encouraged.

(216) Awnings, canopies, and signs that obscure the elements representing the character of the city center should not be used.

(217) Adding awnings and canopies to buildings in the city center shouldn't hurt their important design and architectural features.

(218) Canopies and awnings can be used as signs.

(219) In north-facing streets with low natural light, glass canopies or transparent plastics can be used to provide protection against rain without obstructing natural light.

(220) The lighting under the canopies should illuminate only the sidewalks and facades of shops.

(221) Design guidelines should recommend a suitable combination of canopies and signs, etc., to create a pedestrian scale along the streets of the city center.

7.3.11 *Materials and Colors*

Objectives:

- Improving the quality of individual buildings and the general streetscape of the city center;
- Encouraging the use of high-quality, low-cost, and long-lasting local materials and colors; and
- Reflecting a sense of visual continuity and stability over time.

Design guidelines:

(222) The materials used must be in harmony with the character of the main street.

(223) It is recommended to use local and common local materials and colors to preserve the identity and unity of the street, especially in the building walls.

(224) Materials used in different parts of buildings in the city center, such as those used in the roofs (visible points), external walls of buildings, decorations, and other parts, should be durable and at the same time have a traditional and cultural character as much as possible.

(225) It is recommended to use materials that are attractive in terms of texture and details for roofs and walls. Examples include precast concrete, plasters, EIFS,[1] different types of stone, wood (or materials with wood design), brick, etc.

(226) Mirrors, reflective glass, and stones that reflect light are not recommended for the façade.

(227) New materials can be used in cases where they match the materials used in adjacent buildings in terms of scale, texture, color, and details and complement the building facade design.

[1] Exterior Insulation Finishing System (EIFS) involves the use of a new type of material for building facades that are installed on the external walls by screws. This product prevents water from penetrating the walls and is also used to create beauty and variety in the facade.

(228) Materials such as concrete, stone, and tile can be used on the ground floors, and concrete, stone, tile, wood, and plaster can be used on the upper floors.

(229) The materials of historical and old buildings and artworks should be preserved as they increase the texture quality, integrity, and visual continuity and can improve the character of the streetscape.

(230) The main materials of the facades of historical and old buildings should not be covered or obscured. Where the main building materials have been covered with secondary materials, they should be exposed to the extent possible (Winter & Company 2002).

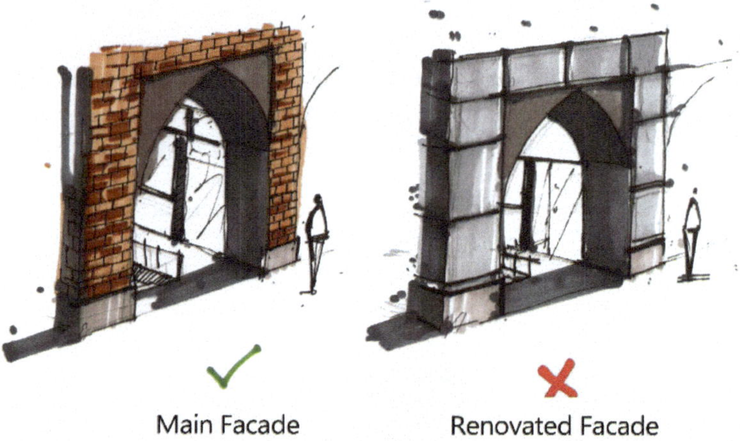

(231) The main walls of the city center must have clean and seamless materials and should be free of all signs, equipment, and mechanical and electrical devices that are no longer used or are destructive to their visual appeal.

(232) If a certain type of building material with a special color (such as red brick) is used in the city center buildings, it is not to use very similar colors for facades and architectural decorations in other buildings (such as the details of the edges of windows, doors, cornices, etc.) (McBee et al. 1992).

(233) In historical urban centers, the similarity in the use of materials should be brought about by a dominant color (for example, the use of thatch-colored materials in the buildings of the historic city center of Yazd or red bricks in the traditional center of Boston).

7.3.12 *Windows*

Objective:

Increasing the continuity and visual appeal of city center streets by creating a human scale.

(234) The main windows of the upper floors should be recreated or made distinctive.
(235) The proportion of modified windows on the upper floors and their arrangement pattern should follow the overall facade design as much as possible.
(236) The ground floor facades should include windows (Nottingham City Council 2006).

(237) Any change in the windows' position or size that affects the architectural rhythm of the facade and the main character of the building is prohibited.
(238) The main shops that attract people should be oriented toward the main street and have display windows or boxes and shelves facing the street.

(239) Display windows should never be blocked or covered, and those that have been changed should be returned to their original dimensions (Paumier 2004).

(240) In harmony with the character of the existing main street, windows should constitute between 10 and 50% of the ground floor facade (Nottingham City Council 2006).

(241) Hidden corners on the street should have at least 50 lx of lighting at night (Pakzad 2013).

(242) The window-to-wall ratio on the upper floors should be less than that on the ground floors.

(243) Floodlights should be semi-transparent, or their glare should be reduced using translucent glass or acrylic lenses.

(244) The use of reflective glass (mirror) in the entire building and reflective and translucent glass in display windows should be avoided.

(245) Where entrances are placed along the same line as display windows, canopies can be used to create diversity and clarity at building entrances.

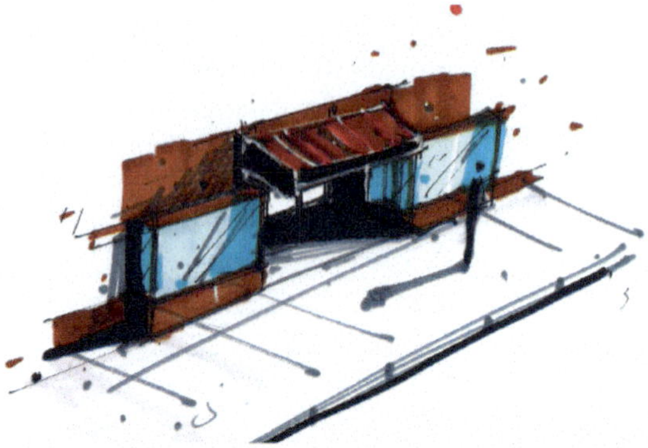

(246) Using multi-piece windows is preferred to integrated ones, because dividing the surface into smaller parts gives the building a more human scale.

7.3.13 Columns

Objective:

Strengthening diversity and visual continuity, creating an integrated architectural framework, and forming a visual focus for the upper facades of buildings in the city center.

Design guidelines:

(247) To strengthen the proportional balance in wide buildings, it is recommended to use columns that separate the facades of shops.

(248) In places where changes have been made to the columns, the width of the columns and their intervals should be modified to increase consistency in the facade (McBee et al. 1992).

(249) To emphasize the role of columns in defining the architectural character of the upper facade, their surface materials should be the same as those in the facade (Paumier 2004).

7.3.14 Signs

Objectives:

- Creating a comprehensive program for signs as well as coherent design criteria for the city center;
- Improving the visual quality of signs through the use of complementary sizes, shapes, colors, and lighting methods;
- Encouraging the use of signs that are both clear and of appropriate scale for their subjects.
- Encouraging and guiding commercial uses that use symbols and graphic designs, creative elements, lighting, and other elements to define their services.

Design guidelines:

(250) Signs must have quality design and materials and be in harmony with the design of the city center.

(251) Signs should be impressive and express their message without additional details or redundant words, but they should not be so abstract as to cause an ambiguous and incomprehensible message (Paumier 2004).

(252) The installation of the following signs should be avoided (LMN 2005b; Paumier 2004).

- Pole-mounted signs;
  Exception: The use of pole-mounted signs related to historical and memorial places with a height of fewer than 5 m with plant landscaping around their base (floor) is not prohibited.
- Signs with uncovered electrical conduits;
- Signs with glaring or moving lights;
- Balancing weights and other equipment;
- Signs with luminous plastic letters;
- Audible-producing signs;
- Thin cardboard signs;
- Temporary/permanent signs installed on rooftops (such as banners, etc.);

(253) Signs can be installed on the surface of buildings as long as they do not hide or obscure their historical and identity-inducing details.

(254) Signs should be coordinated with the design of buildings and other signs in terms of scale and size. This requires the preparation of a coordinated plan for signs in the city center.

(255) Coordinated signs should be used to strengthen the visual quality and human scale of spaces.

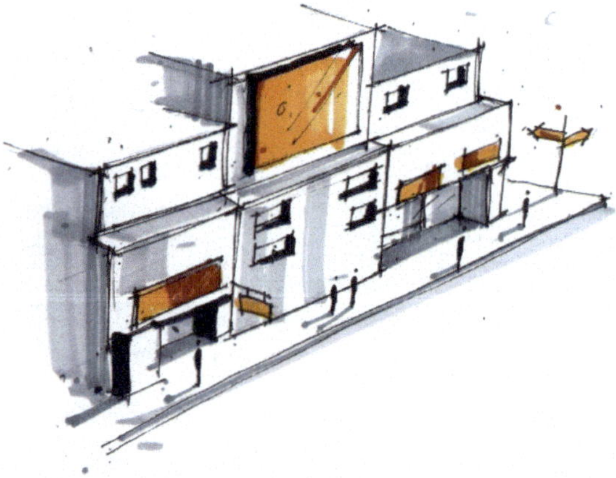

(256) Blade signs that are hung from or mounted on canopies, arches, or shop fronts can be used to increase visibility.

(257) Blade signs should not have an area larger than 2800 cm^2, and their thickness should not exceed 5 cm. Also, these signs should be at least 2.4 m above ground level (LMN 2005b).

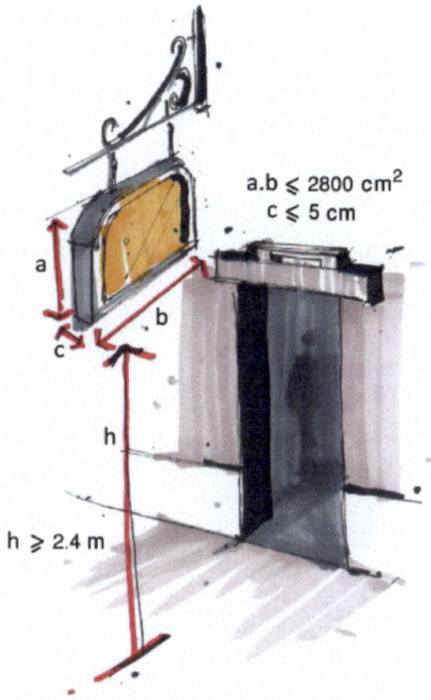

(258) The ratio of one square foot (about 930 cm^2) of sign area for every foot (48.30 cm) of the front width of the building is a good guide for the dimensions of the signs on building facades (Paumier 2004).

(259) The surface of building walls and facades should be considered as part of the general design of signs.

- The overall composition of the facade, including its decorative details, should be taken into account.

- Signs should be appropriate to the building so as not to overshadow its appearance.
- Signs should be in harmony with the facade in terms of scale.
- Signs should be installed on the building in such a way that they can emphasize the design components of the facade itself. In no way should the sign obscure the details or architectural features of the building.

(260) The following points should be considered when placing and installing signs:

- Signs should strengthen the horizontal lines of the borders and facades along the street.

- Signs should not obscure the architectural elements and characteristics of buildings. For example, wall-mounted signs should not hide the columns of a building.

(261) The tenants on the upper floors can (in line with the city center's coordinated sign plan) use signs above the facades of shops on the ground floor and under

the cornices or roof lines. In most cases, wall signs should be above the display windows and under the canopy, with an approximate height of 5.2 m above the ground (Winter & Company 2002).

(262) Wall signs should be placed in line with other signs in the block if possible.

(263) Attempts should be made to include a minimum decorative border for wall signs (so that it does not cause ambiguity in the important features and elements of the facade). The borders of the facades and other building decorations and details (especially in centers with historical backgrounds) can also be used to provide a proper background and frame for the signs (Winter & Company 2002).

(264) Signs can include graphic shapes and volumes to emphasize their intended type of activity and product.

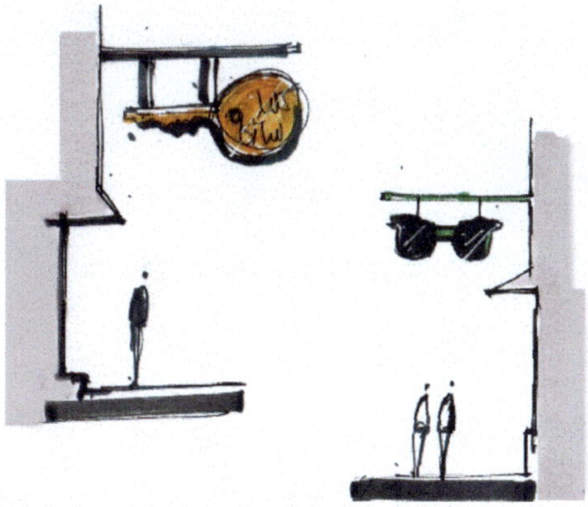

(265) Where several commercial activities use a single building, their signs should be coordinated. To this end, several small guide signs can be grouped in a row within a frame. Similar shapes and backgrounds should be used to make them easy to read while connecting them visually.

(266) In urban centers with historical backgrounds, signs that do not match the character of the buildings and the details and decorations of the facades should be avoided.

(267) The signs that are visible at the eye level of passers-by can be painted on the front glass of shops.

(268) The independent standing signs should not interfere with the function of traffic and pedestrian signs or their visibility.

(269) Information and guide signs should be placed at stops and intersections to inform citizens and tourists.

(270) Neon can be used artistically in signs. However, neon should not be simply used to highlight and distinguish the roofs or edges of buildings (Paumier 2004).

(271) Signs that are flashing or dazzling, vary in the amount and concentration of light, or move should be avoided (McBee et al. 1992).

(272) Signs should have an indirect light source if possible (Paumier 2004).

7.3.15 Setbacks

(A) **Upper-floor setbacks**

Objective:

Using upper floor setbacks to reduce the volume of tall buildings, strengthen pedestrian scale in the street facade, and create a reasonable level of visibility of the sky from the street.

Design guidelines:

(273) The street-facing facades of upper floors exceeding 7.5–20 m in height (depending on the scale of the city center) are required to have a minimum 3-m setback. Clearly, this slope should increase in proportion to the height (LMN 2005a).

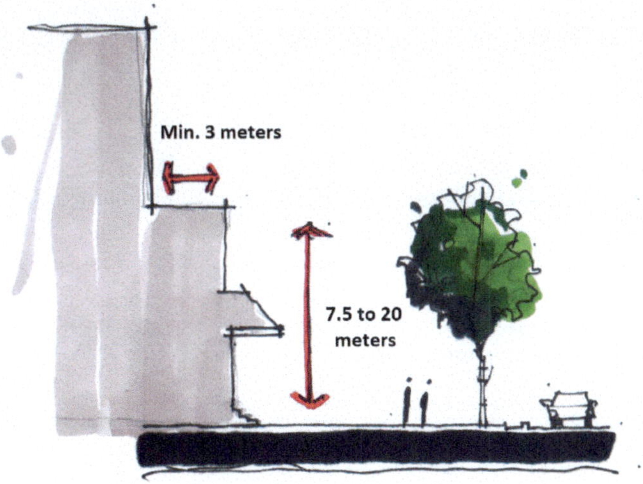

(274) Visual Angular Plane Analysis can be used to calculate the amount of setback for taller buildings. However, this method should be used in association with sun/shade analysis, three-dimensional modeling of the studied space, etc. (Mcllroy 2006)

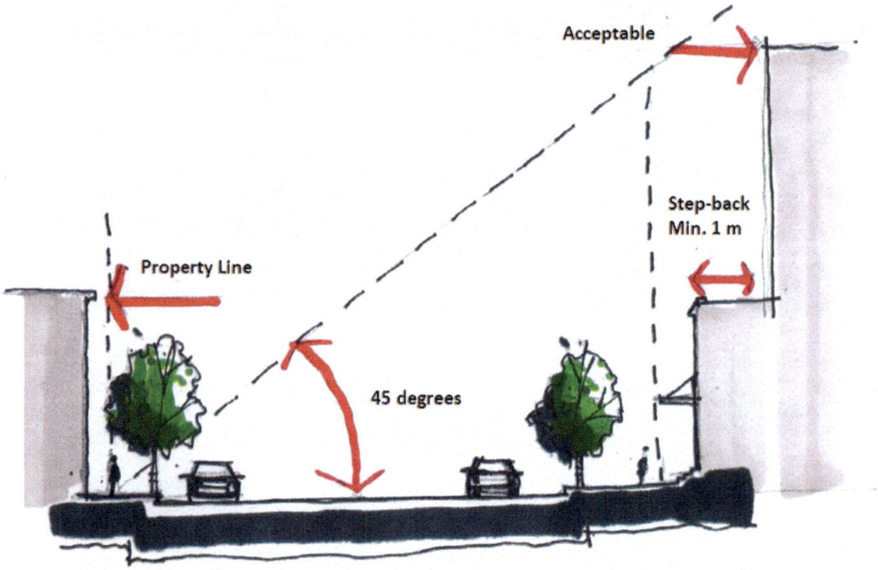

Visual Angular Plane

(275) If there is a requirement to create a tall building in the city center, the higher floors should be set back from the main facade, and the lower floors should be located along the edge created by existing buildings.

(276) The upper floor setback can be done in the form of a sloping roof.

(277) The visual focal points of buildings do not need to be set back (LMN 2005a).

(B) **Front setbacks**

Objective:

Creating a continuous and harmonious building edge and humanizing the pedestrian environment.

Design guidelines:

(278) New buildings must follow the setback of existing buildings so that the building edges are continuous and coordinated.

(279) In the central core of the cities, existing setbacks should generally be adapted to pedestrian movement routes, and active uses at the ground floor of the street should be able to open directly to the pedestrian area.

(280) The setback of buildings with side yards should repeat the space rhythm of existing buildings (Paumier 2004).

(281) The setbacks of the front and side yards of buildings should be reduced as they approach the city center to be effective in creating a compact urban development pattern (Paumier 2004).

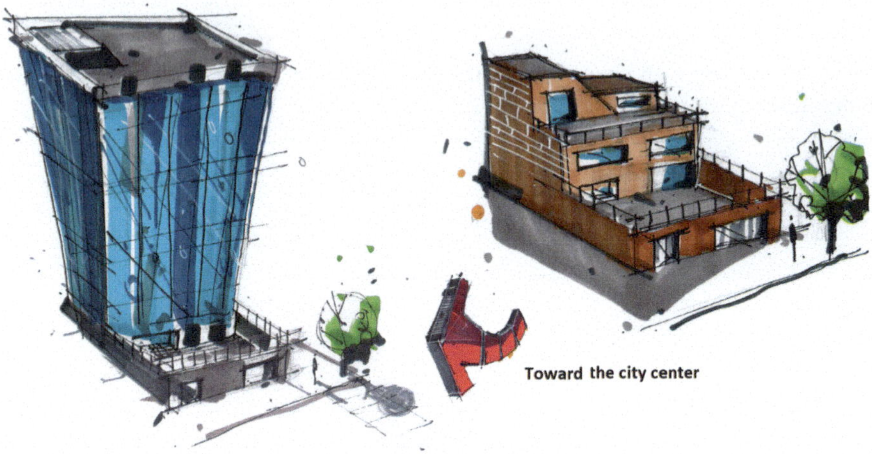

(282) The setbacks of the street front should be deep enough to create wide and landscaped sidewalks to humanize the pedestrian environment.

7.3.16 Rooftop Landscaping

Objective:

Creating a skyline with visual appeal and integrating the design character in the city center through rooftop design.

Design guidelines:

(283) High-rise buildings should take measures to make their roof form unique and yet articulated. These can include the following options:

- Recesses related to the terraces
- Raised cornices
- Grids along the parapet
- Geometric volumes (dome, globe, etc.)
- Material or color changes in the upper floors
- Decorative frieze
- Decorative inscriptions (capitals).

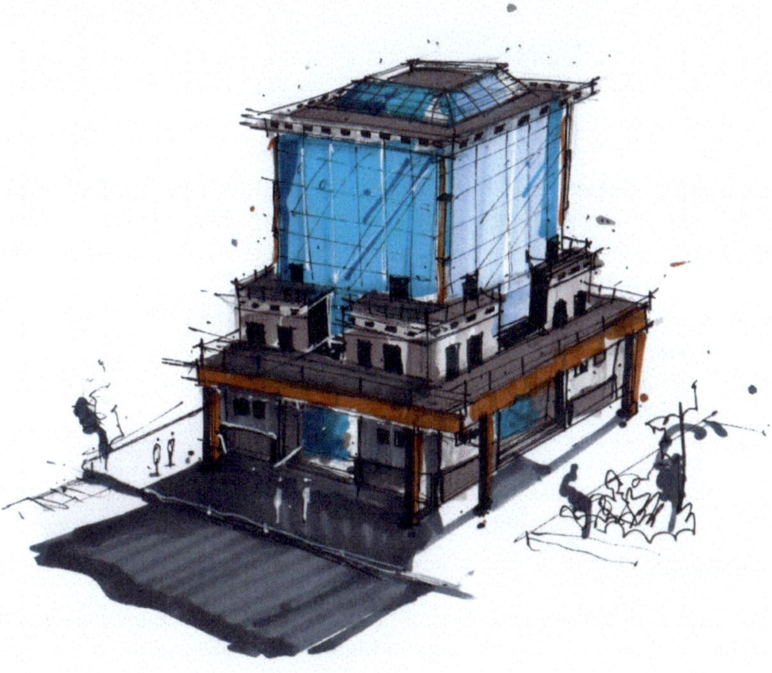

(284) Unifying components should be used in the design of city center roofs. Therefore, if possible, it is recommended that buildings have integrity in the slope, elements, details, materials, and configuration (overall volume) of roofs.

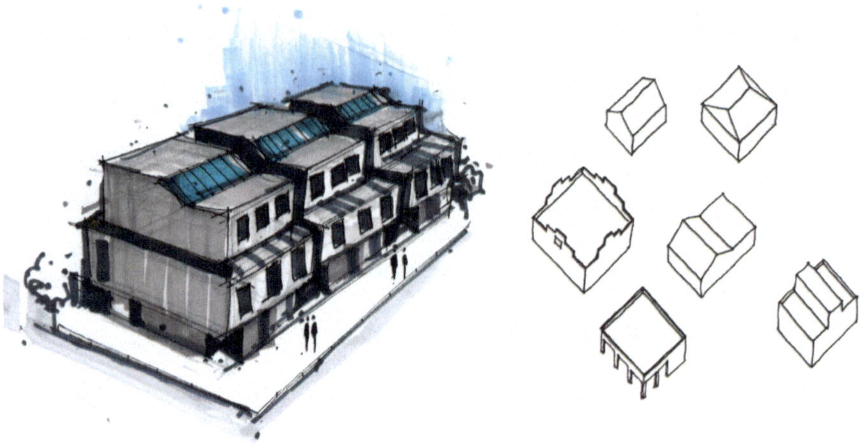

(285) The equipment and mechanical facilities behind the parapet should not be visible to pedestrians in the city center.

(286) All the mechanical and service equipment on the roofs must be placed inside the opaque walls of the parapets or behind (or inside) the shapes of the roofs. Whatever method is used for creating the barrier, must be integrated into the architectural design of the building.

7.3.17 Elements and Focal Points

Objective:

Creating distinctive elements and focal points to direct the sight of passers-by and strengthen the sense of gathering around them.

Design guidelines:

(287) Any distinctive element and focal point, including towers, domes, minarets, statues, clock eye-catching parts in buildings, must have the prior approval of the city center design commission.

(288) The location of these points should be appropriate, providing a visual permeability from the city center's main streets.

(289) Attempts should be made to maintain the visual sequence of signs and distinctive elements.

(290) Visually valuable focal points of buildings do not have to be set back.

7.3.18 *Additions to Buildings*

Objective:

Reducing the possible negative effects of new buildings annexed to the old and historical ones in the city center.

Design guidelines:

(291) The additions should be proportionate and harmonious with the main building in terms of scale, materials, and character.

(292) The additions should not obscure or damage the important architectural or historical features of the first building.

(293) An addition that induces a time period before the construction of the main building should be avoided as it can create a false sense of history.

(294) The added parts must be distinguished from the old building with due elegance. Even if this is done through minor and subtle differences, the main character of the building should remain interpretable.

(295) Added buildings and parts can be placed behind or next to a building, provided that they do not create a false sense of history and can arrange the facade components of the shops, borders, cornices, plinths, and windows of the upper floors in the same way as in the older building (Winter & Company 2002).

(296) The new part can be added to the roof of the old building, provided that it is located behind the primary façade that defines the character of the old building to help preserve the historical scale of the building's appearance (Winter & Company 2002).

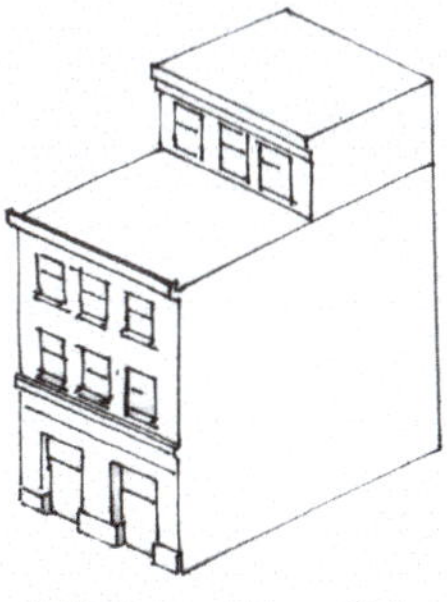

The new building has been added as a roof to the setback part of the main three-floor building.

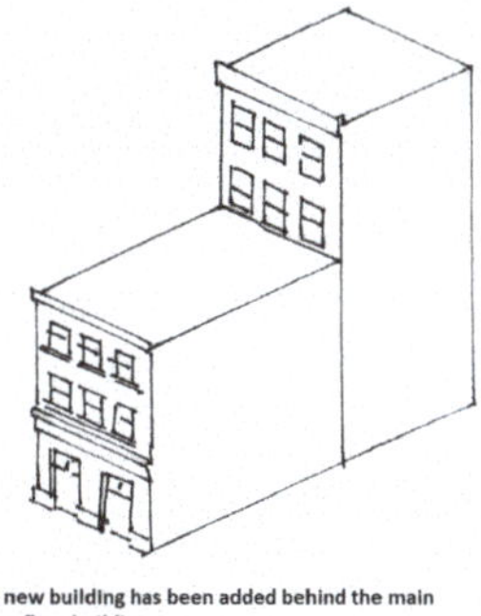

The new building has been added behind the main three-floor building.

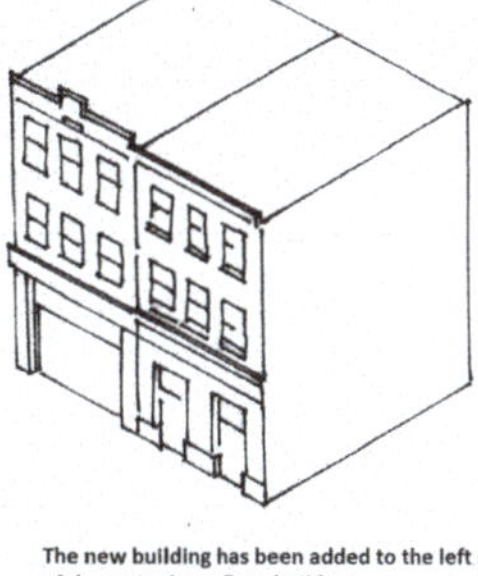

The new building has been added to the left side of the main three-floor building.

7.3.19 Back and Side Facades of Buildings

Objectives:

- Achieving the favorable impact of these facades on the mental image and visual character of the city center
- Maintaining visual continuity and presenting an image of a city center block

(A) **Corner buildings**

Design guidelines:

(297) The facades of the shops in such buildings should be rotated at the corner so that street view consistency and visual continuity are not disrupted.

(298) In this category of buildings, the corner facade design should primarily follow the facade facing the street and show the transition from one area of the city center to another (Paumier 2004).

(299) Design guidelines used for the improvement of shop facades and the upper floors of the main front should also be used for corner fronts.

(B) **Unfinished corner facades**

Design guidelines:

(300) Where unfinished facades are visible from the street, their quality and view should be improved by removing or covering the exposed mechanical equipment and expanding the use of materials, colors, or details similar to those in the front walls of the buildings.
(301) Where windows cannot be created, graphic designs can be effective in making walls attractive. Such graphics should be designed with a neutral color background (Paumier 2004).

(C) **Back facades**

Design guidelines:

(302) Back facades should be attractive and create an identity related to the front facade of the buildings.
(303) All surfaces of the walls must be well restored.

(304) Service areas and warehouses should be carefully organized and placed out of the sight of pedestrians.

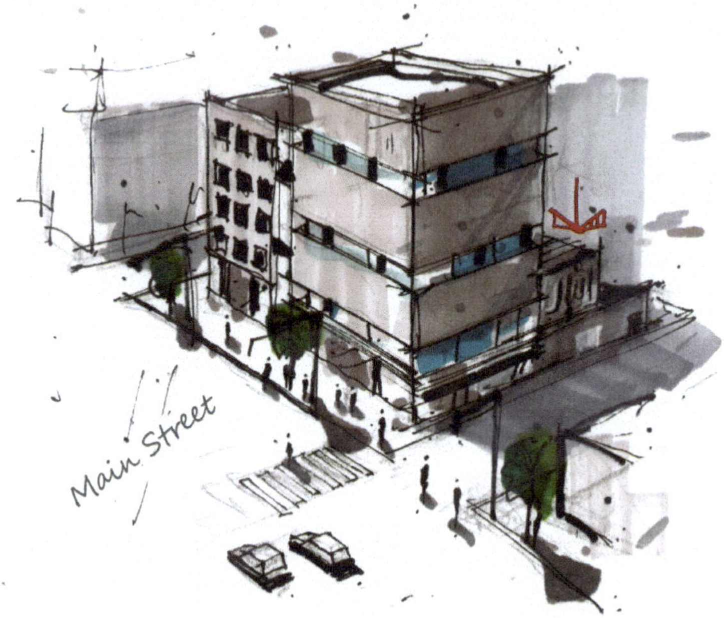

(305) Entrance doors, commercial signs, and lights should add to the external appeal of the facade.

(306) Awnings, display windows, and landscape design can also strengthen the back facades.

(307) Blocked windows must be opened again.

(308) The back entrances of buildings should be used for service purposes, storage, and access to parking lots and should be upgraded to a safe space with sufficient lighting and guide signs to provide access to parking lots for customers and employees.

7.3.20 *Uses and Functions*

Design guidelines:

(309) Uses that generate noise and visual pollution (auto repair and car dent repair shops, parking, etc.) should not be located in city centers. Even if they cannot be removed, they should not be located in the first spatial layer of the street but should be transferred to the spatial back layers as concentrated complexes.

(310) The establishment of large-scale, inactive uses during non-business hours within the city center's street should be avoided.

(311) Uses that require relaxation (hospitals, universities, kindergartens, etc.) should not be located at the front of the main street in urban centers.

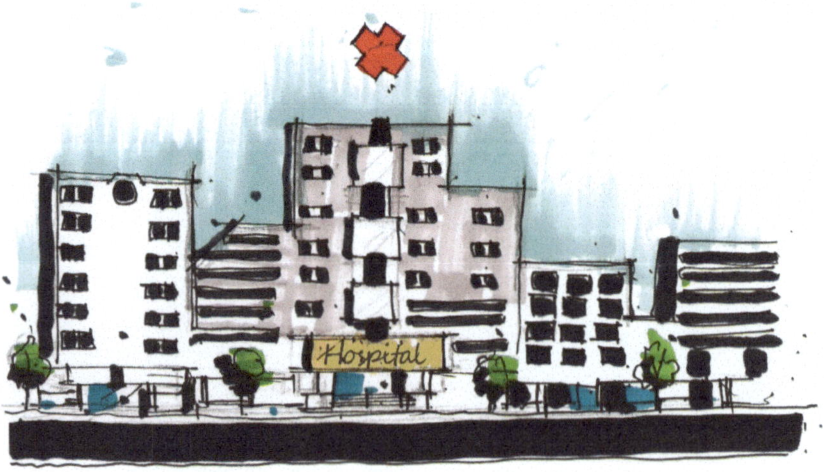

(312) Outdoor cafes should not impede the movement of pedestrians or vehicles. At least a 1-m setback from the edge of the street curb is required in areas where street cafes are established (Community Development Department 2003).

- To facilitate the movement of pedestrians, the width of the sidewalk must be free and unused for at least 2 m (Gould Evans Goodman Associates 2005).
- Places equipped for street cafes should not block the movement of pedestrians into and out of nearby commercial properties.

(313) The body of uses such as street cafes, which need to place furniture in the outdoor space, should have setbacks on the ground floor front.

(314) It is recommended to provide facilities for activities that bring vitality and increase the presence of people (such as street vendors, carnivals, festivals, etc.).

(315) It is mandatory to establish round-the-clock functions and uses on the main street.

(316) There should be a maximum distance of 250 m between nighttime uses and activities (restaurants, theaters, etc.) (Pakzad 2013).

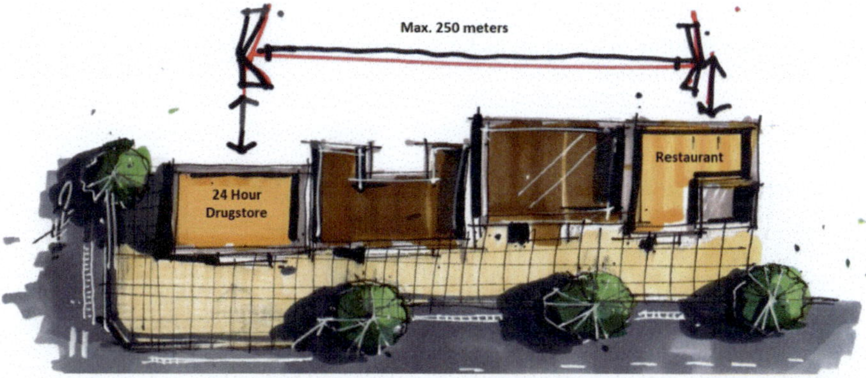

(317) Converting residential and commercial units in the body of the city center main street and surrounding neighborhoods into warehouses is prohibited.

(318) Commercial units adjacent to the main streets of the city center should not be abandoned for more than 2 years (Pakzad 2013).

(319) Commercial buildings located on the edges of the main streets should have entrances, windows, and signs facing the sidewalks of the street or the main pedestrian pathways.

(320) The commercial part of residential-commercial units must face the main street. Their residential part can have an access road from the front, back, or side. In any case, it should have a view of the main street as much as possible.

(321) Uses and spaces that encourage people to pause and stop should be encouraged along the street.

(322) Pedestrians from nearby residential or commercial areas should have easy access to the city center. To this end, dense construction is recommended, as it connects access roads and parking lots so that intra-city shopping trips can be combined with pedestrian activity, thereby reducing the number of unnecessary roadways and traffic loads on city streets.

(323) Small shops or services are used as part of the main street front to create pause points. To meet this need, it is necessary to include a setback of 90–175 cm (Community Development Department 2003).

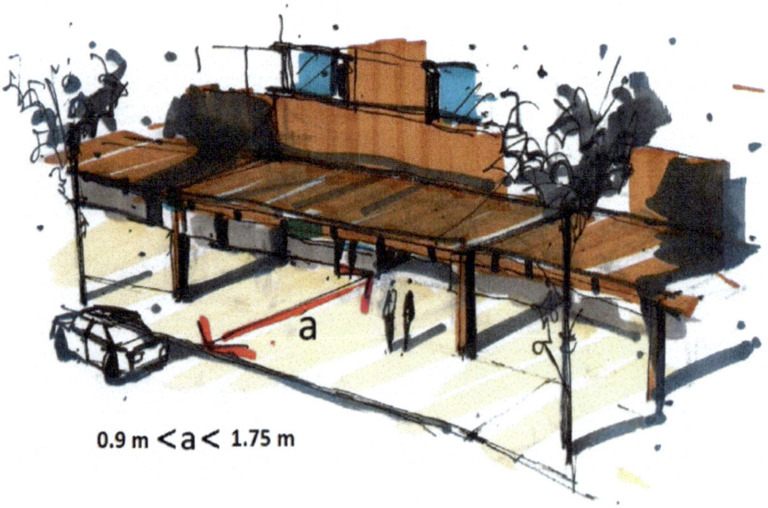

(324) In order to ensure a sufficient height on the ground floor of buildings for effective commercial operation, the height of the first floor for new buildings on the main street should not be less than 3.5 m and should be proportionate to the surrounding structures (Nottingham City Council 2006).

7.3.21 New Developments

(A) **Infill developments**

Objective:

Eliminating empty spaces created by vacant lots and open parking lots, strengthening continuity in the urban physical environment, and promoting functional and visual continuity in the city center.

Design guidelines:

(325) New constructions should be located in such a way as to increase activities along the street edge and in front of the building.

(326) Infill buildings should be carefully designed to strengthen the positive features of the existing architecture.

(327) To ensure coordination with existing structures, infill developments should use key elements of existing constructions in a compatible, but not imitative, manner.

(328) Where existing buildings are poor in terms of visual richness and design standards, repeating their design components will not be desirable, and new buildings should be used to formulate new contexts and standards.

(329) The ratio of building height to width should be maintained in infill developments.

(B) **New Large-Scale Developments**

Objective:

Integration and optimal combination of large and high-rise buildings with the existing context of the city center.

Design guidelines:

(330) The dominant scale and setbacks of existing buildings should create a framework within which the new architecture fits well.

(331) Incongruous forms and volumes should be avoided in new buildings.

(332) New high-density buildings should be created considering easy walking distances with concentrated traditional business units (less than 9 min of walking) (Paumier 2004).

(333) There should be transitions between the height and volume of large projects and the existing or proposed structures in their neighborhood.

(334) Mandatory height-to-setback regulations can reduce the visibility of tall buildings from old street fronts. In addition, horizontal and vertical setbacks and similarity with surrounding buildings on the lower floors can help to reduce the negative visual effects of such buildings (McIlroy 2006).

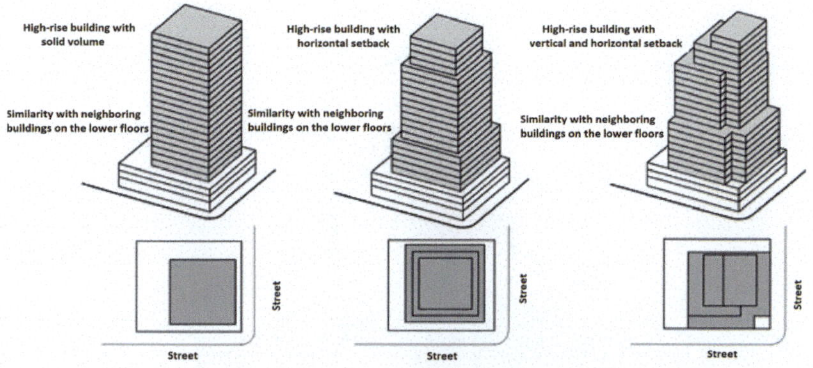

(335) In high-rise buildings with setbacks on the upper floors, facade details (such as the lines of cornices, plinths, openings, etc.) on the floors of the same height as adjacent buildings should be similar to the details used in the facades of surrounding buildings and not clutter the visual rhythm and continuity.

(336) Large structures should be designed in such a way as to maintain pedestrian communication and visual corridors and share their internal activities with the street.

(337) The construction of buildings that turn their backs to the city center and place continuous blank walls in the pedestrians' field of view, or that direct all their activities to the inside should be avoided in the city center (McBee et al. 1992).

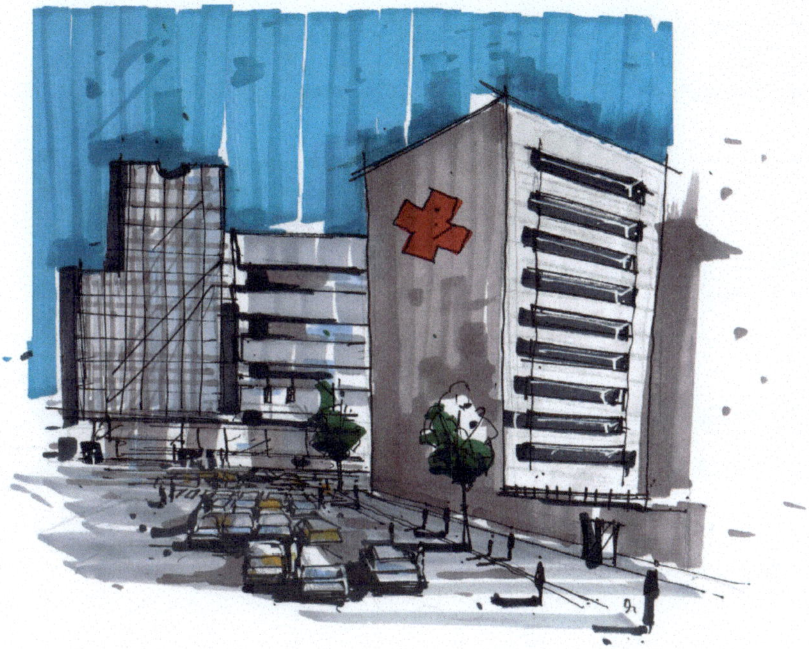

(338) As mentioned in the design guidelines for infill developments, new buildings must be able to properly complete the blank facades of existing walls. However, in conditions where the infill development is large-scale, articulation measures should be used to reduce its adverse effects.

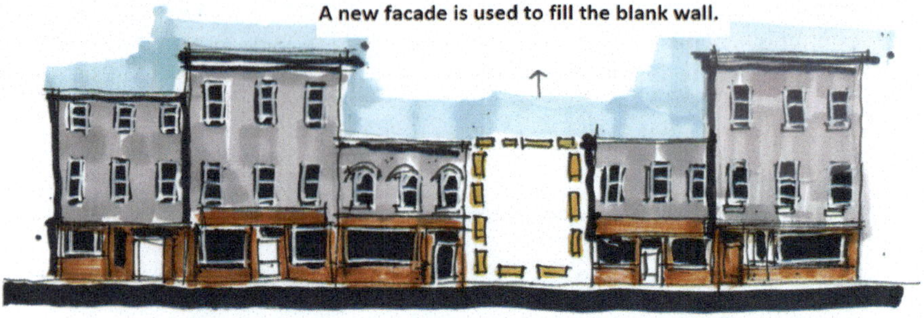

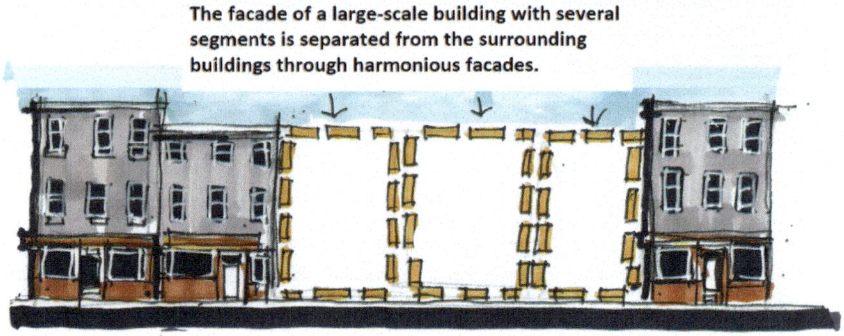

(339) Major facades and entrances of large-scale buildings should be oriented toward the city center's main streets.

(340) Commercial and retail uses on the ground floors of large-scale buildings can ensure visual and functional continuity in the city center (Paumier 2004).

7.4 Design Guidelines for Access System in City Centers

7.4.1 Roadway

Design guidelines:

(341) Car speed should be reduced in the active areas of the main streets and other busy districts in the city center to increase pedestrian safety.

(342) The maximum and minimum number of lanes for each direction of a roadway are 3 and 2, respectively. These lanes should not be wider than the usual width (3 m), because in that case, the speed of cars will increase to an unacceptable level.

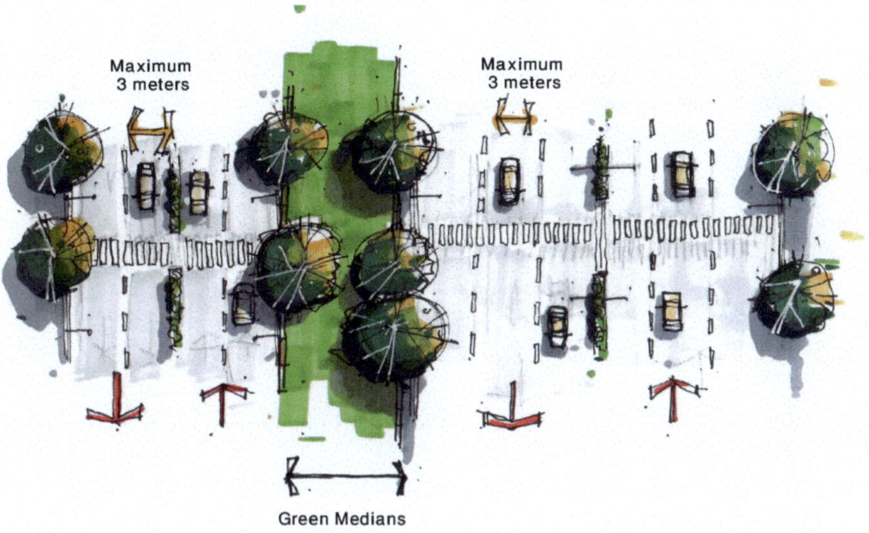

(343) If the main street is wider than allowed, green medians and pedestrian refuges or buffer spaces can be created. Otherwise, it is necessary to maintain the existing usable width of the street.

(344) The speed of cars in cities can be slowed down by using different pavement materials, like stone, on roads that are wider than allowed.

(345) The necessary facilities for public transportation vehicles should be provided so that they are prioritized over other means of transportation.

(346) The maximum distance between bus stops on the main street should be considered 800 m (Pakzad 2013).

(347) 300 m is the minimum distance between intersections on the city center's main streets (Pakzad 2013).

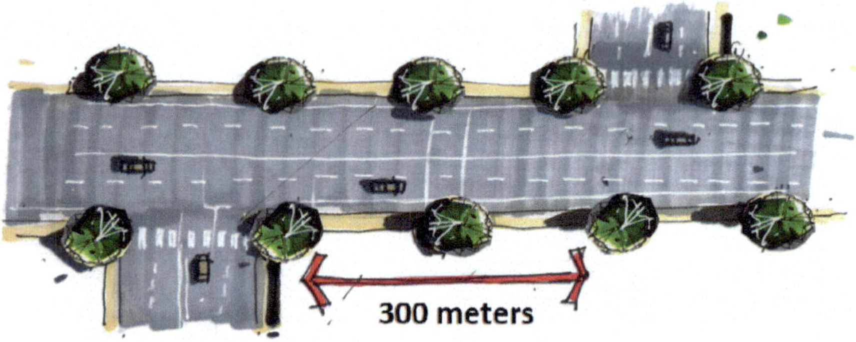

(348) Complex intersections with more than four branches are prohibited (Pakzad 2013).

(349) At the intersection of the main street and the local street, some elements can be used to induce a gateway.

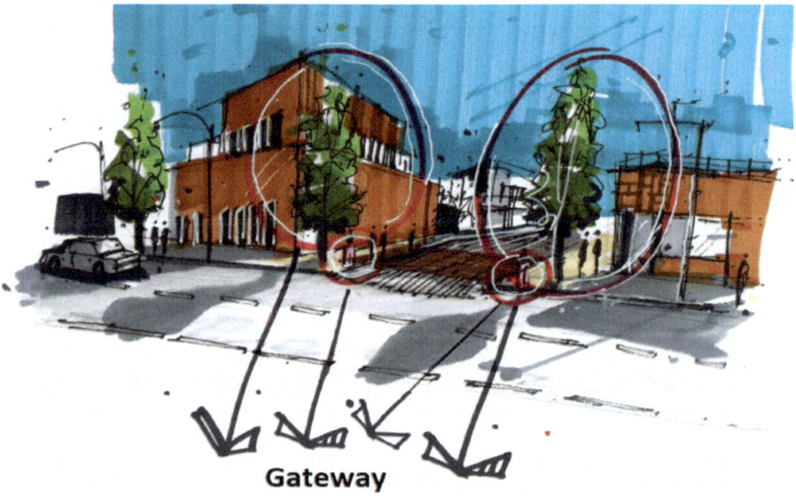

(350) If an interchange is needed, constructing an underpass perpendicular to the main street is allowed.

(351) The construction of any type of car overpass perpendicular to or along the main street is prohibited.

(352) Important main streets should not be one-way, even if possible.

(353) The intersections of the side streets with the main street should be rectangle if possible.

7.4.2 *Sidewalks*

Design guidelines:

(354) Sidewalks should be wide enough to accommodate a large number of pedestrians, be equipped with sales stalls, bus stops, street trees, and street furniture, and connect various activities and functions.

(355) The proper width of sidewalks in the major spines and main streets of the city center is 6 m, with approximately 3.6 m allocated to the sidewalk and the area adjacent to storefronts in order to facilitate the movement of the passing crowd that may stop to observe stores. The area between the curb or street edge and a distance of 2.4 m shall be designated for furniture, public artworks, and the presence of street cafes (Paumier 2004).

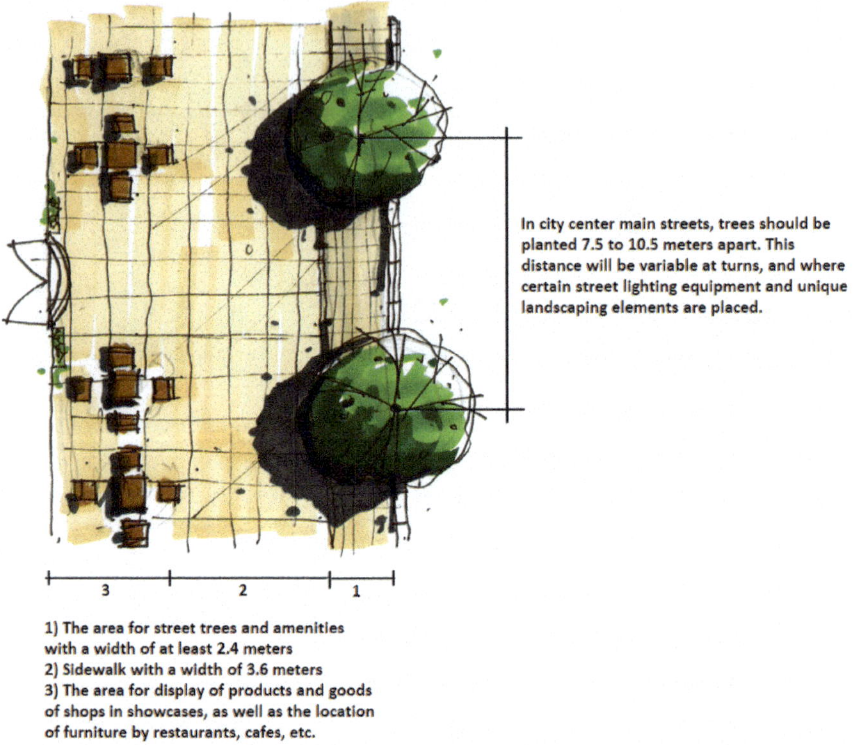

(356) Wide sidewalks if used in small cities, appear empty and lose their dynamism. Therefore, in such urban centers, sidewalks with a width of more than 4.3–4.9 m are not recommended (Paumier 2004).

(357) Sidewalks in the city center's main streets should have an acceptable variety in terms of construction and finishing of their components and materials (e.g., pavers, colored concrete, etc.).

(358) Sidewalks should help to create a sense of visual integration and improve pedestrian experiences on main streets.
(359) Physical barriers on the sidewalk edge (fences, wooden rails, vertical buffers, etc.) that separate cars and pedestrians should be removed.

(360) Flat gardens should be dedicated to planting and creating green spaces in the amenities area of the sidewalk in accordance with the street tree planting program.

(361) Asphalt and concrete should not be used for sidewalk pavement.
(362) The sidewalk pavement should be capable of getting reconstructed and repaired easily and quickly.
(363) The texture of the sidewalk pavement must be completely anti-slip.

(364) Near intersections, the sidewalk pavement should be turned into a strip to inform the blind of reaching the intersection.

(365) Uncovered tunnels are allowed only in places that have permanent flowing water.

(366) Allocating proper spaces for parking bikes and motorcycles on the main street sidewalks can reduce disturbance and clutter.

(367) If the sidewalk slope is close to the maximum allowed amount, level landings should be created at certain intervals.

(368) If the conditions permit, private open spaces can be incorporated into the landscape design of the public sidewalk so that they expand along the same line. Design a plaza with wide sidewalks and extend it into a building's private courtyard, for instance, to enhance the visual effects of these spaces.

(369) Placing a single step or a sudden variation in walking surfaces along the sidewalk is prohibited.
(370) Building stairs and entrances should not encroach on the sidewalk.

(371) If the ramp or stairs of existing buildings have to be placed on the sidewalk, they should be placed and used along the building and not perpendicular to it. In this regard, setbacks are recommended to accommodate the stairs.

(372) At crosswalks in busy streets, variation in the surface can be exploited if possible to further reduce the speed of cars

(373) Bus stations should not encroach on the useful width of the sidewalk.

(374) At crosswalks, the pavement or its design should be altered. At crosswalks, if possible, the edge of the sidewalk can protrude slightly into the roadway to slow down automobiles.

(375) The lighting of the sidewalk should be in harmony with the lighting of valuable walls and buildings.

(376) Fine-grained materials should be used for ramps and around water-related landscaping elements (drinkers, fountains, etc.).

(377) Surface and subsurface drainage measures, including a suitable slope, should be taken for sidewalk pavement.

(378) Walkways between buildings and parking lots should be enhanced with lighting, green spaces, planters (e.g., climbing or hanging plants on walls, bases, etc.), and the addition of color, texture, and attractions. In the city center, the use of flowers and plants, as well as lighting, can create a sense of security, continuity, and attractiveness.

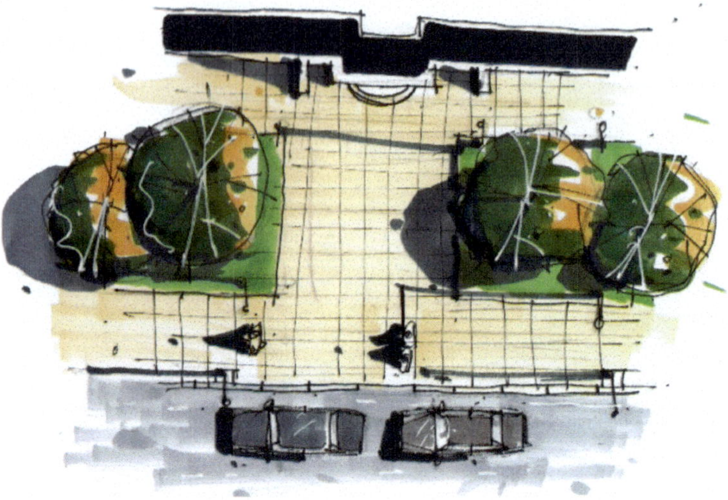

(379) Creating a protrusion in the sidewalk and reducing the width of the street at a distance of at least 50 m from the intersection can prevent the unauthorized stopping of cars and make it easier for pedestrians to cross the street (Pakzad 2013).

(380) Sidewalks in the main streets of the city center can be designed as roofed walkways.

(381) In order to improve the experience of pedestrians walking between buildings or between cars or public transportation vehicles and commercial buildings, a continuous and comprehensive network of sidewalks should be created to reduce collisions between pedestrians and cars. In new construction as well as in existing commercial structures, a clear, continuous, and protected network of sidewalks must connect the following:

- Main entrances of adjacent buildings located in the same area;
- Entrances of key buildings and parking lots in the city center area; and
- Main entrances of buildings overlooking the sidewalks.

(382) The continuity of pedestrian and bicycle movement should be maintained at intersections.

7.4.3 *Corners and Setbacks in City Center Streets*

Design guidelines:

(383) The corners of all main street sidewalks should be aligned along the street.
(384) The corners should be emphasized by variation in form, skyline, and colors or materials.

(385) It is necessary to create ramps in all corners and cuts of the sidewalk to facilitate access and movement for all user groups, including the disabled.

(386) The density of plants and street furniture should be reduced at the corners to provide a free and safe field of view.

(387) Spaced with a free and safe field of view at the corners are desirable and attractive places for pedestrians. Therefore, the necessary equipment should be provided for pedestrians to stop for a short time there, but this should not lead to disturbance and crowding in these areas.

(388) In order to define the boundaries of the street and improve the level of activity and attractiveness, buildings should have a minimum amount of setback (Community Development Department 2003).

(389) The setbacks of the facades facing the main street should not exceed 175 cm (Community Development Department 2003).

(390) The main street landscape should continue a little along the lower-level streets connected to it at the corners so that it gradually ends by merging into its surrounding area.

(391) Setback and removal of a portion of the building mass on the corner is an acceptable method for facilitating pedestrian movement, increasing driver visibility, and identifying corner outbuildings.

(392) Part of the front facades of buildings can be set back in order to provide the space needed to create plazas, street cafes, and landscaped areas.

7.4.4 Parking

Objective:

Ensuring that parking is not a noticeable and dominant element in the city center.

Design guidelines:

(393) Parking should be avoided at intersections.
(394) It is better to place parking behind buildings (with a secondary access road connected to the city center main street). As a second resort, they can be placed on the sides of buildings (English Partnerships 2006).

(395) Continuous pedestrian paths must be created inside the parking (in order to connect all milestones and activity points for pedestrians).

(396) The existing connections for pedestrian traffic between parking and major activity centers along the central border and the commercial backbone of the city center should be attractive and appropriate. For this purpose, the private sector builders can help by creating access roads within the block (the distance between the parking and the city center) with active uses embedded in their edges (Paumier 2004).

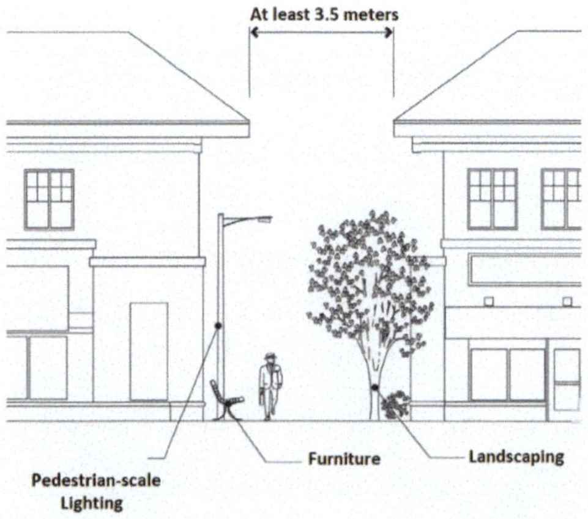

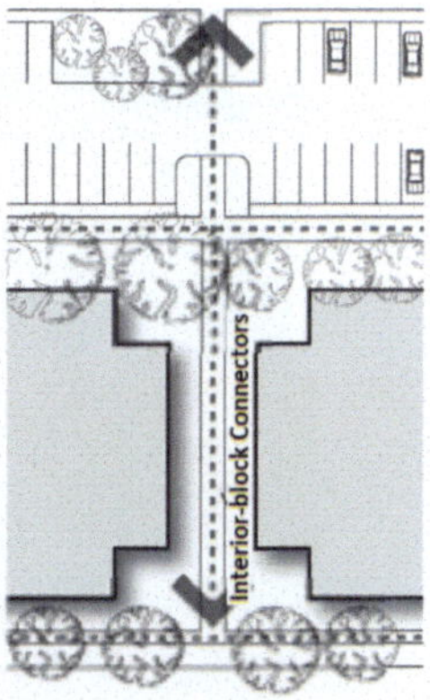

(397) The size of the parking should be coordinated with the requirements of existing buildings.
(398) Wherever possible, parking spaces should be placed on the lower floors of buildings or within interior-block structures behind active uses. Otherwise, they should be placed next to the buildings.
(399) Open parking lots should not create large distances in the development structure or separate the important and key uses of the city center from each other.

(400) In cases where on-street parking is needed, the parallel type is preferred to the angle one (Paumier 2004).

Objective:

Landscaping should be used to eliminate the visual effects of open parking areas and increase green space in the city center.

Design guidelines:

(401) A landscaped environment with plantings and shade trees should soften and keep out of sight the harsh environments of parking lots.

(402) Landscaping of parking lots should be used to enhance the flow of pedestrians and cars. This landscaping should be done for:

- parking lot entry points;
- the endpoints of the car corridors; and
- walkways in parking lots.

(403) Landscaped sloping surfaces should not be used to separate surfaces and create barriers in the city center (Paumier 2004).

(404) In the landscaped areas close to the sidewalks, a green area with a height of 1.2–1.5 m above the ground should be created using plants or hedges (LMN 2005a).

(405) Open parking lots should be divided into separate and clearly defined units using plants, footpaths, special seating areas, or other landscaping elements.

(406) The mature height of trees in parking lots should range from 70 to 140 cm (Community Development Department 2003).

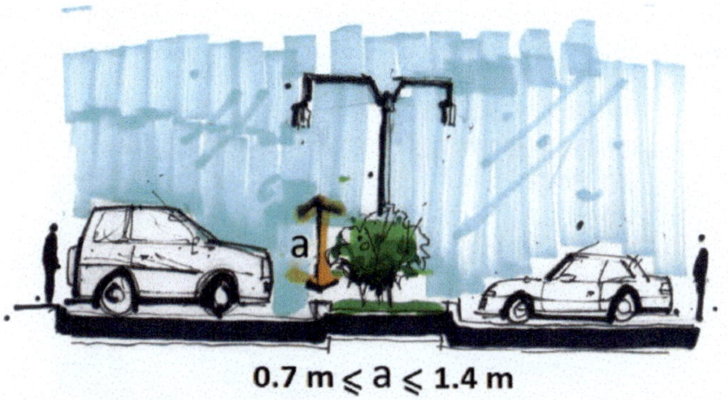

(407) While asphalt, concrete, or impervious surfaces are the common materials used for open parking spaces, other permeable surfaces such as porous paving, concrete, green paving (a combination of paving and grass), etc. can also be used for this purpose.

(408) The number of trees required in the interior landscaped area of parking lots depends on the location of the parking lot in relation to the building and public right-of-way: (LMN 2005b)

- Where the parking lot is located between the building and the public right-of-way, one tree for every four spaces shall be enough (1:4).
- Where the parking lot is located in the side yard and borders the public right-of-way, one tree for every six spaces shall be enough (1:6).

- Where the parking lot is located in the rear yard and is not visible from the public right-of-way, one tree for every eight spaces shall be enough (1:8).

Objective:

To reduce the impact of open parking lots on the streetscape by creating buffers.

Design guidelines:

(409) In parking lots tangent to the walkway, a buffer should be created by one or a combination of the following measures (LMN 2005b; Paumier 2004):

- Short concrete or masonry walls with a maximum height of 1 m at the edge of the parking lot;
- Planter walls with at least 50% evergreen plants and a maximum height of 1 m (climbing and vertical plants are preferred);
- Landscaping using plants that include 50% evergreen trees, shrubs, and other vegetation growing on the ground.

(410) Landscaped edges with plant elements should be created with a width of at least 1.5 m between parking lots and adjacent developments to reduce their visibility for pedestrians.

(411) Trees planted in the green belt at the edge of the parking lot should not have a distance of more than 3 m from the crown center of each other. Also, at least one bush should be planted for every 1 square meter of a landscaped green area (Community Development Department 2003)

(412) If the parking space is tangent to a plot of land, a buffer strip of landscaped space with a width of at least 1.2 m should be created between the parking or its paved surfaces and the property lines tangent to it (Department Development Community 2003).

(413) The buffer should not weaken the safety of the environment.

(414) The part of the buffer that is higher than 1 m in height must allow more than 75% visibility from the outside through measures such as metal fences (Paumier 2004).

(415) Chain fences should not be used for creating buffers or blocking or closing parking lots along the city center sidewalks.

Objective:

Reducing the visual impact of parking structures in the city center.

Design guidelines:

(416) Parking structures and on-street parking lots should be used as an alternative to surface parking areas.

(417) The design guidelines for street-level uses, transparency, protection against weather, and details should also be implemented on the ground floors of these buildings.

(418) If these parking lots are located in the main pedestrian-oriented streets of the city center, at least 50% of the space in front of them on the ground floor should be allocated to commercial spaces (McBee et al. 1992).

(419) The parking structure facades that face the street should be similar to other normal floors and avoid the creation of open spaces full of cars and roof lights visible from the outside. This is done by implementing one or two of the following: (English Partnerships 2006):

- Square openings instead of horizontal ones;
- Planters for flowers and other plants on the façade;
- Decorative metal grids;
- Frosted glass; and
- Murals or public arts, etc.

(420) Parking structures should be designed in such a way as to prevent the creation of distinct plots.
(421) These parking structures should achieve a finer-grained human scale through segmentation of facades, recesses, and placement of the first floor below the existing level line, reducing the overall height.
(422) Parking structures should be enclosed by commercial or office uses outside to blend in with other existing buildings along the city center street sidewalks. Where this is not possible, the visual impact of such parking should be enhanced by natural or artificial landscaping.
(423) Parking structures should be designed in such a way as to improve the dynamics of the street scene in the city center.
(424) If possible, a parking structure in the city center can be covered by a multi-story shopping-commercial space so that it is not visible from the street level and the whole building will be visually attractive in this combination (Winter & Company 2002).
(425) If parking structures are created in a city center with historical background, they should follow the architectural patterns and traditional elements and details used in the facades of older buildings (e.g., windows, cornices, etc.).

(426) In other streets where the ground floor of parking structures is occupied by parked cars, a landscaped setback should be used to soften their visual impact on the street and sidewalk.

(427) The length of a parking structure on each street should not exceed 76–91 m, and its width should not exceed 55 m (Paumier 2004).

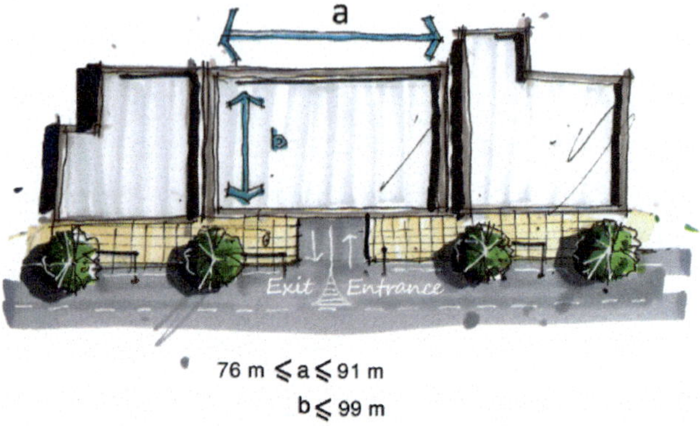

(428) In sensitive places (for example, near residential areas), neighborhood areas should be protected from the lights of parking (Winter & Company 2002).

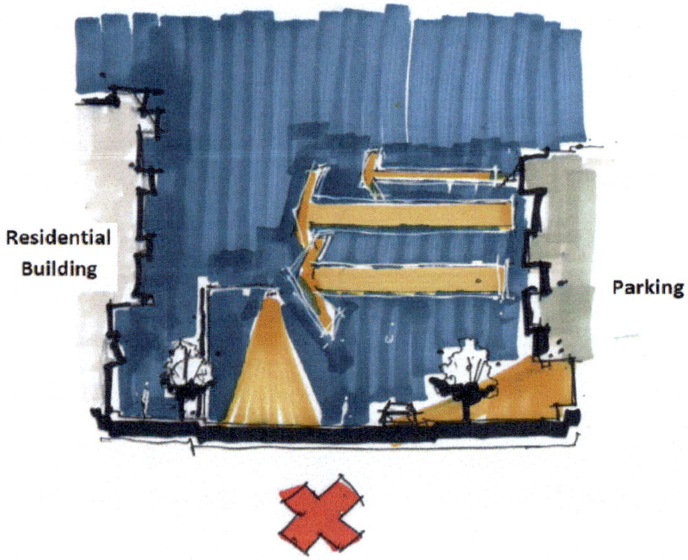

(429) Corner parking structures are preferable to other types (Paumier 2004).

(430) Materials used in the architecture of parking structures should be similar to those used in the exteriors of nearby buildings.
(431) The window patterns of adjacent buildings should be repeated on the front facade of the parking structures.
(432) Neutral architectural colors and forms (rather than strange shapes, materials, or colors that attract attention) should be used for parking structures.
(433) Underground parking are preferable to parking structures in terms of their visual effects on the city center.

(434) Separate individual parking should use the design guidelines related to the details of the ground floor facades as well as those for awnings and guards.

Objectives:

Creating parking lots in the central area will encourage people to walk in the city center and also ensure the visual continuity of the street front.

Design guidelines:

(435) Open parking lots should not be located in front of commercial buildings along the city center's main streets.

(436) Wherever possible, it is better to reduce the parking structure façade along the main street and create a sense of enclosure by placing small structures such as kiosks or market stalls at the parking entrance.

(437) Public parking structures should not occupy more than one commercial opening along the street wall.

(438) High-visit user should have their own parking spaces, which should be located on the back layer or basement.

(439) The entrances of public parking structures should be 100 m away from intersections (Pakzad 2013).

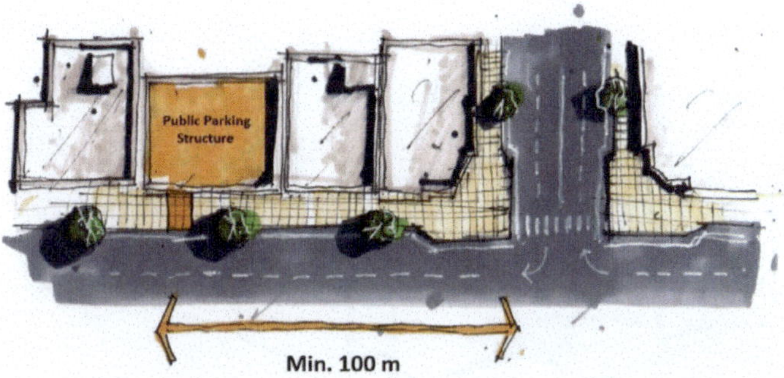

(440) If open parking lots have more than 200 parking spaces, they should be divided into smaller sections to reduce the visual impact of parking to a minimum. Plant landscaping with a width of at least 2.4 m can be used to divide large parking lots (LMN 2005b).

(441) Curb cuts created for parking lots on the sidewalks of commercial areas in the city center should be minimized. This can be done through common entrances and exits (if possible) (English Partnerships 2006).

(442) Parking lots should be established in such a way as to reduce the blank walls along the continuous street edge to a minimum.

(443) Where parking lots interrupt the pattern of facades on the street, the entrance causing such an interruption must have a minimum width.

Objectives:

Considering pedestrian safety and security in relation to parking lots.

Design guidelines:

(444) A forecourt at the entrance can be created to distinguish the parking and pedestrian spaces and increase pedestrian safety. This forecourt should have an area of at least 20 square meters and be created with a height of about 15 cm above the parking level to create a visual-spatial distinction (LMN 2005b).

(445) All parking lots should include 1.5-m-wide walkways (marked by color or pavement type) to connect the parking lots to the sidewalks adjacent to building entrances (English Partnerships 2006).

(446) The sidewalks within parking lots must be elevated above the ground and have an edge or other barriers to prevent cars from entering their territories.

(447) Walkways inside the parking (structured parking or parking lot) should be visible and free of closed and hidden spaces.

(448) The lighting used in parking lots and their landscaped areas should create visibility and security and should not exceed 7.5 m in height. Moreover, their dazzling brightness in the surrounding areas should be reduced as much as possible (Paumier 2004).

(449) The level of parking lots should, if possible, be lower than the surrounding ground level with a height difference of about 1.2–2.4 m so that their areas can be seen from the street level (McBee et al. 1992).

(450) Mixed-use design should be expanded in activity nodes near the entrance of the parking.

References

Community Development Department (2003) City of Sumner urban design concept plan: Design and development guidelines. Sumner, Washington
DMAP (2004) Downtown streetscape plan for city and county of Denver. Denver
English Partnerships (2006) Car parking: what works where toolkit. Retrieved from http://www.englishpartnerships.co.uk/carparking.htm/
Gould Evans Goodman Associates (2005) Destination downtown overland park design guidelines. Kansas
LDR International Inc. (1998). City of Columbia, South Carolina city center design/development guidelines. Columbia
Lenexa Planning & Development Department (Lenexa) (2007) Lenexa City center neighborhood, design standards& guidelines. Lenexa, Kansas
LMN Architects (2005a) City center design guidelines. City of Lynnwood, Washington
LMN Architects. (2005b) Mill Creek town center design guidelines. Mill Creek, Washington, DC
McBee S, Urban Land Institute, Basile J, Ralph, Urban Land Institute Urban Development/Mixed-Use Council, Petersen C, David L, Phillips P (1992) Downtown development handbook. ULI-the Urban Land Institute, Washington, DC
McIlroy B (2006) Downtown urban design guidelines. City of Burlington.
Nottingham City Council (2006) Nottingham city centre streetscape design manual. Nottingham
Pakzad J (2013) Urban space design guidelines in Iran. Ministry of Housing and Urban Development Press
Paumier C (2004) Creating a vibrant city center: urban design and regeneration principles. Urban Land Institute
Winter & Company, Hoshide Williams Architects, Nakano Associates, LLC (2002) City center design standards. City of Bellingham, Bellingham, Washington

Uncited References

Aleksandrowicz FJ (1973) The arcade: enclosed shopping area in downtown Cleveland. Retrieved from http://www.tomclarkblog.blogspot.com/2012/01/first-mall.html
Allposters (2012) Procession in the Forum Romanum, Ancient Rome. Retrieved from http://www.allposters.com/-sp/Procession-in-the-Forum-Romanum-Ancient-Rome-Posters_i4230402_.htm
Arbury J (2010) Time to humanise the CBD? Retrieved from http://transportblog.co.nz/wp-content/uploads/2010/09/usa-trip-050910-022.jpg
Bahreini SH (2016) Urban design in Iran, a new approach. Fine Arts Magazine, 26
Basirat M (2008) City center. J Urban Manag 17
Behzadfar M (2016) Urban development plans and programs: Concepts, trends and requirements of comprehensive and detailed plans in Iran. Shahr Press, Tehran, Iran
Benevolo L (1977) History of modern architecture. MIT Press
Bentley I (1985) Responsive environments: a manual for designers. Routledge
Calthorpe P (2005) From new regionalism to the urban network: changing the paradigm of growth. Harv Des Mag 22:60–66
Carmona M, Oc T, Tiesdell S, Heath T (2003) Public places—urban spaces. Architectural Press
Caves RW (2005) Encyclopedia of the city. Routledge
Cinema Treasures (n.d.) Historic theater [photograph]. Retrieved from https://cinematreasures.org/theaters/767/photos/197063

Cparama (2012) Cours la Reine. Retrieved from http://www.cparama.com/forum/rouen-le-cours-la-reine-t7708.html

Geoffsholler (2013) The Greeks. Retrieved from http://www.geoffsholler.com/the-greeks.html

Golkar K (2008) Constructive components of urban design quality. SOFEH J 11(32)

Howard E (1965) Garden cities of to-morrow, vol 23. MIT Press

Internet Archive Book Images (n.d.) Historical illustration from "The British Flower Garden" [Illustration]. Flickr. https://www.flickr.com/photos/internetarchivebookimages/14597954080

Jacobs J (1992) The death and life of great American cities. Vintage Press

Lang J (1994) Urban design: the American experience. John Wiley & Sons

Murgaš F, Klobučník M (2018) Quality of life in the city, quality of urban life or well-being in the city: conceptualization and case study. Ekológia 37(2):183–200

N (n.d.) Map of Milan tram: tram lines and tram stations of Milan. Retrieved July 1, 2022, from https://milanmap360.com/milan-tram-map

National Archives and Records Administration (NARA) (n.d.) The Arcade, enclosed shopping area in downtown Cleveland [Photograph]. Wikimedia Commons. Retrieved from https://commons.wikimedia.org/wiki/File:THE_ARCADE-ENCLOSED_SHOPPING_AREA_IN_DOWNTOWN_CLEVELAND_-_NARA_-_550286.tif

Nipic (2011) Belgium: most popular European tourist destination. Retrieved from http://www.nipic.com/show/1/73/4667701k638299c3.html

Pakzad S (2016) Revision of the design of a memorial urban square (Master's thesis). Iran University of Science and Technology

Proshansky HM (1978) The city and the self identity. In: Environment and behaviour. Cambridge University Press, pp 97–118

Schürer R (2012) Notre Dame de Paris von vorne. Retrieved from http://www.mint.sbg.ac.at:8000/pics/2008/Paris/Notre%20Dame/index_low.html

Shakibamanesh A, Behzadfar M (2009) The location of design guidelines in urban design process and their role in promoting urban spaces quality. Armanshahr Architecture Urban Develop J 1(1)

Shakibamanesh A, Behzadfar M (2011) Promoting city center parking qualities by using urban design guidelines. Int J Architect Urban Plann 1(1)

Space (2011) Dynamic spaces in European cities. Retrieved from http://www.space-eu.info/space_conferences.aspx?confContentId=197

Schultz MS, Kasen VL (1985) Encyclopedia of community planning and environmental management. Facts on File Press

Spragg NZ, Bernard (2019) Downtown Toronto. Retrieved October 26, 2022, from https://www.flickr.com/photos/volvob12b/49663498897

Tallon AR, Bromley RD (2004) Exploring the attractions of city centre living: evidence and policy implications in British cities. Geoforum 35(6):771–787

Tavasoli M (1997) Urban design in the central part of Tehran. Iran Urban Planning and Architecture Studies and Research Center Press, Tehran

Tavasoli M, Bonyadi N (2010) Urban space design, vol 1. Iran Urban Planning and Architecture Studies and Research Center Press, Tehran

Terabass (2012) Times square, New York [Photograph]. Wikipedia. https://en.wikipedia.org/wiki/File:New_york_times_square-terabass.jpg

Turkzadeh N (2017) Urban space; elements and criteria (Master's Thesis of Urban Design). Shahid Beheshti University, Tehran, Iran

Von Sternberg M (n.d.) Brent civic centre. Retrieved October 26, 2022, from https://www.skanska.co.uk/expertise/projects/56952/Brent-Civic-Centre

White HW (1980) The social life of small urban space. The Conservation Foundation. Washington, DC

White HW (1989) Cities: rediscovering the center. Doubleday, New York

Wikimedia Commons (n.d.) Drawing of Rue Notre-Dame, Montreal, 1843 [Drawing]. Retrieved from https://commons.wikimedia.org/wiki/File:Rue_Notre-Dame_Montreal_1843.jpg
Wikimedia Commons (n.d.) Naqsh-e Jahan Square [Drawing]. Retrieved from https://commons.wikimedia.org/wiki/File:Naghse_jahan_-_drawing.jpg
Wikipedia (2014) Tehran Grand Bazaar. Retrieved September 25, 2021, from https://upload.wikimedia.org/wikipedia/commons/8/8b/Tehran_Bazaar_old.jpg

The manufacturer's authorised representative in the EU is Springer Nature Customer Service Centre GmbH, Europaplatz 3, 69115 Heidelberg, Germany. If you have any concerns regarding our products, please contact ProductSafety@springernature.com

Printed and bound by CPI Group (UK) Ltd, Croydon, CR0 4YY
01/10/2026
02225926-0014